JIGS AND FIXTURES

MAXWELL PRESS

JIGS AND FIXTURES

Fred H. Colvin

Associate Editor, American Machinist;
Member A.S.M.E.

&

Lucian L. Haas

Tool Designer; Member A.S.M.E.

MAXWELL PRESS

Chennai Trichy New Delhi

MAXWELL PRESS

This edition has been published in india by arrangement with Carson Books, UK

ISBN 978-93-5528-208-8 **Maxwell Press**

No. 44, Nallathambi Street,
Triplicane, Chennai 600 005

MJP 1291

Publisher : C. Janarthanan

Publisher's Note

The legacy of a country is in its varied cultural heritage, historical literature, developments in the field of economy and science. The top nations in the world are competing in the field of science, economy and literature. This vast legacy has to be conserved and documented so that it can be bestowed to the future generation. The knowledge of this legacy is slowly getting perished in the present generation due to lack of documentation.

Keeping this in mind, the concern with retrospective acquiring of rare books has been accented recently by the burgeoning reprint industry. Maxwell Press is gratified to retrieve the rare collections with a view to bring back those books that were landmarks in their time.

In this effort, a series of rare books would be republished under the banner, "Maxwell Press". The books in the reprint series have been carefully selected for their contemporary usefulness as well as their historical importance within the intellectual. We reconstruct the book with slight enhancements made for better presentation, without affecting the contents of the original edition.

Most of the works selected for republishing covers a huge range of subjects, from history to anthropology. We believe this reprint edition will be a service to the numerous researchers and practitioners active in this fascinating field. We allow readers to experience the wonder of peering into a scholarly work of the highest order and seminal significance.

MAXWELL PRESS

PREFACE

In these days of modern manufacturing, jigs and fixtures have become necessary both for interchangeability and reduced costs. And, while every tool designer is confronted by different problems, there are certain fundamental principles which can be modified to meet existing conditions and applied to most cases.

Bearing this in mind, we have endeavored to show these principles, as nearly in the order of their application as possible, so that the designer can select such parts and methods as seem best suited to his particular problem. While we have not by any means exhausted all the known devices used in jig work, we have endeavored to show enough to enable a suitable choice to be made and to suggest other methods should these not meet all requirements.

THE AUTHORS.

September, 1913.

CONTENTS

CHAPTER I

SYSTEM IN THE TOOL ROOM

One of the first essentials of the tool-designing department is that it be run systematically, and the following simple method is in successful operation by some of our largest manufacturing companies.

The part is carefully drawn full size and the jig laid out. The

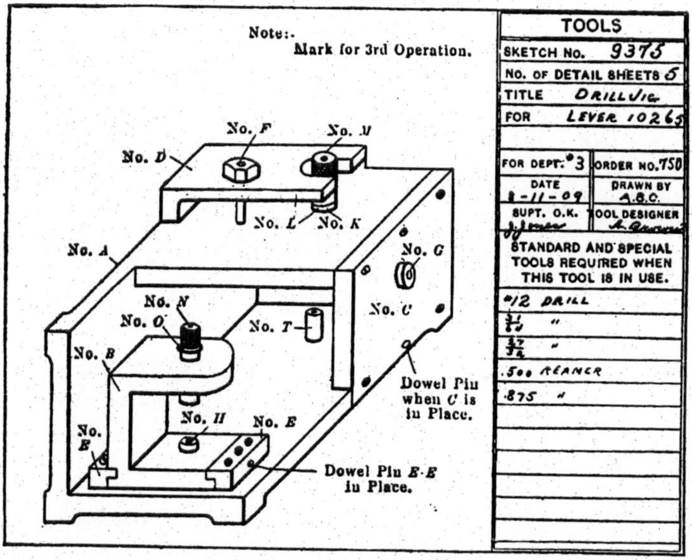

FIG. 1. SKETCH OF COMPLETE JIG.

drawing is then inspected by the chief tool designer and after meeting his approval the sketch is made as follows:

(*a*) A perspective view is made as shown in Fig. 1, sketches being made by the use of three sheets of carbon paper underneath. One of these is a tissue paper so that blueprints can be made from it. The other carbon copy goes to the toolmaker and the original copy is retained for the file in the tool-designing department.

(*b*) The next operation is to make a pencil and one carbon copy only, of all detail parts to be finished on milling machine, shaper, lathe, etc. All forging or special parts are then placed again on separate sheets as shown in Figs. 2, 3, 4, 5 and 6.

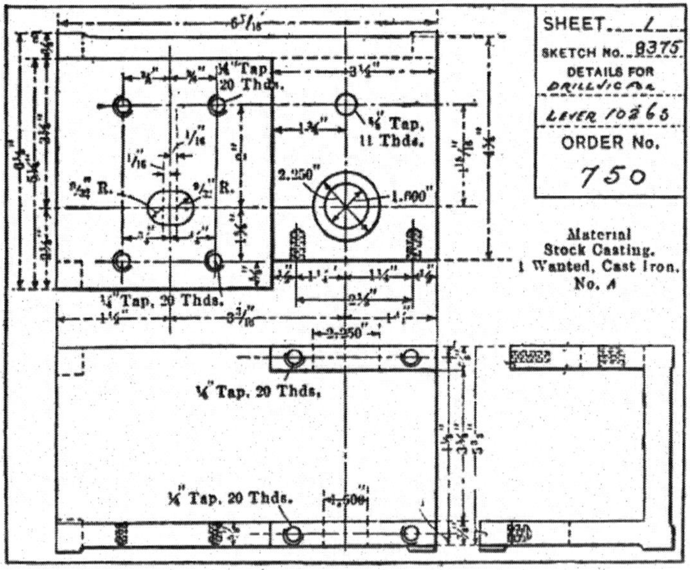

FIG. 2. DETAILS OF BODY.

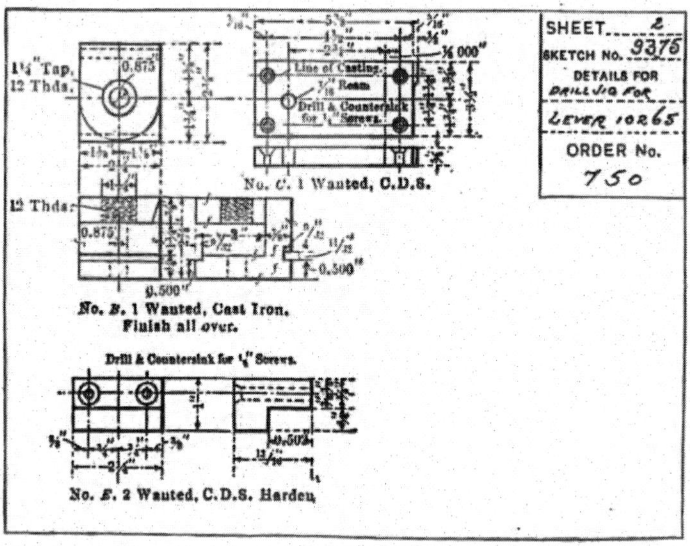

FIG. 3. DETAILS OF SLIDE.

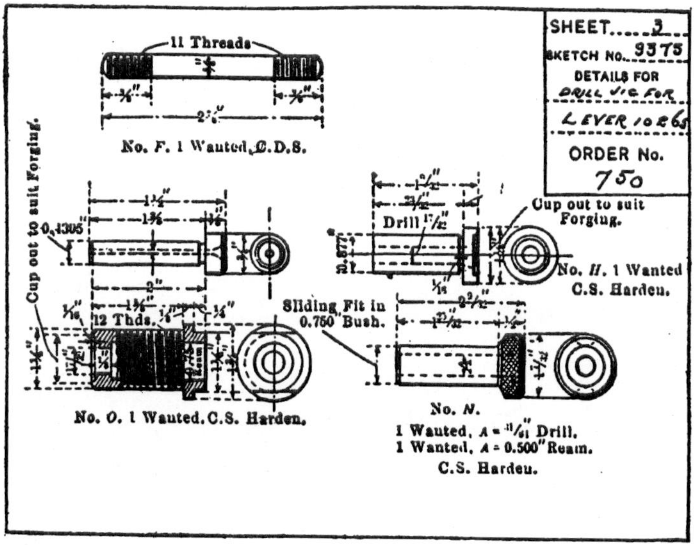

FIG. 4. SLIDES AND BUSHINGS.

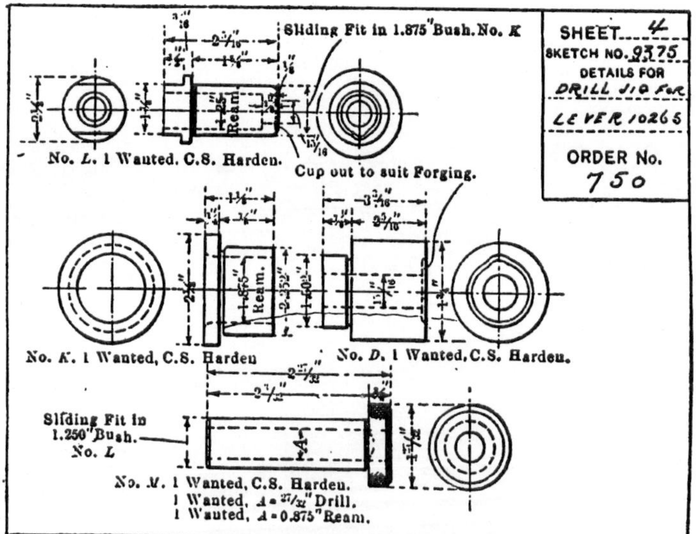

FIG. 5. THE OTHER BUSHINGS.

(*c*) All of these sheets are given the tool sketch number and entered in the tool record. Then everything is checked and the number of the operation for which the tool is made is placed on the perspective sheet.

(*d*) A blueprint is made from the tissue copy of the perspective sketch and sent to the tool store or tool crib so that the man in charge always has a means of knowing just what the tool looks like. This is extremely convenient in case a tool has been accidentally put in the wrong bin or misplaced, as the sketch gives a much better clue than any description possibly could.

(*e*) Every sketch must be O. K.'d by the chief tool designer, the tool-room foreman, and the superintendent, after which they can be sent to the cost department for orders.

(*f*) When the orders are received they are given a coat of shellac and then turned over to the tool room and the tool is made.

TOOL DESIGNING DEPARTMENT'S ACCESSORIES

The numbering of the tool record book and tool sketches can be done with a duplicate numbering stamp.

The tool sketches are made on an ordinary 8 x 10 in. scratch

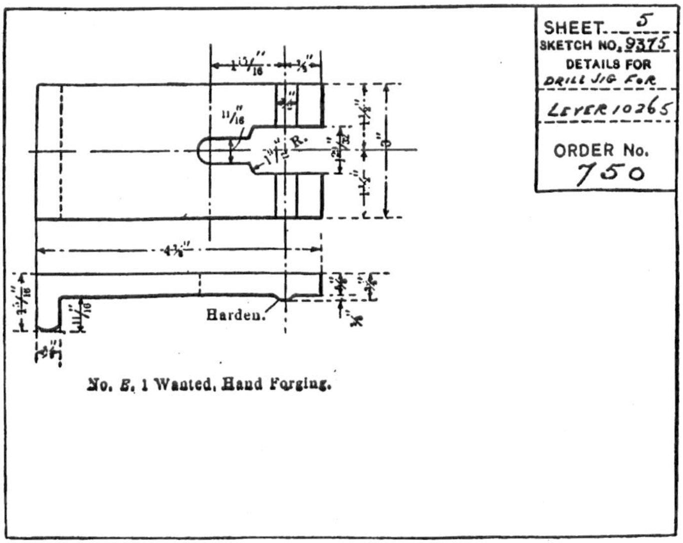

FIG. 6. SKETCH FOR BLACKSMITH.

pad and a rubber stamp for the heading of the perspective sheet can be used as shown on the sketch in Fig. 1 and for the detail sheets as shown in Fig. 2.

In conjunction with this, a series of rubber stamps will be found very useful and helps to give the sketch a neat appearance in case any note such as VOID,–RUSH,–HOLD-UP,–TEMPORARY, etc., is desired to be added to it.

In conjunction with the sketching, a good drafting machine is very necessary to assist the draftsman in producing quick work.

With the above system it is remarkable how quickly a sketch can be turned over to the toolmaker. It often happens a temporary jig, arbor, pin gage or some other tool which is wanted in a hurry does not require a layout but simply a sketch giving the toolmaker the necessary dimensions and that important point of having a record of the tool.

TOOL RECORD AND OPERATION SHEETS

The operation that each piece of work undergoes should be decided in this department and is then entered upon the operation sheet as shown in Fig. 10. The sketch is stamped accordingly with the sketch number taken from the tool record and the number of the operation taken from the operation sheet.

The tool record consists of a well-made book with the numbers running consecutively as shown in Fig. 7. The sketch gets its number from this book, after which the tool is entered on a card as shown in Fig. 9 and this card is placed in a reference file as shown in Fig. 8.

The tool record simply acts as a check in taking out numbers and prevents a number being taken out twice.

The reference card is filled under the part number and is consulted in case one wishes to know what tools are made up for that particular part.

There are several styles of operation sheets or cards and these are generally made to suit the condition governing the factory. Fig. 10 is an example of an operation sheet which can be used to advantage almost under all conditions. It is made of tracing cloth and the operations, etc., are written upon it with a soft pencil, the reason for doing this being to quickly take care of any changes or additions which may arise and the sheet being made of tracing cloth facilitates the making of an indefinite number of prints which can

FIG. 7. TOOL-RECORD BOOK.

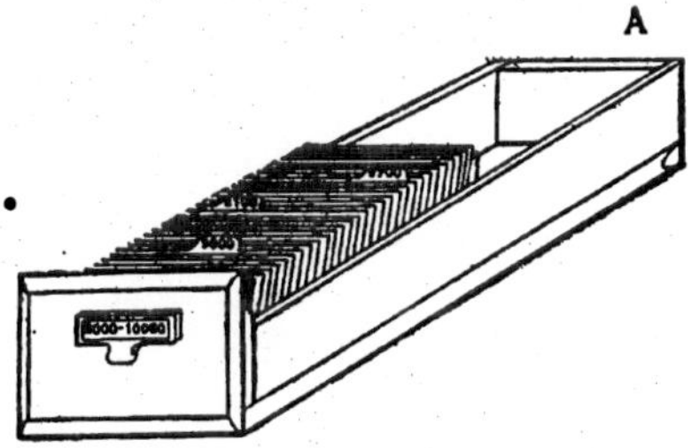

FIG. 8. CARD REFERENCE FILE.

LEVER	10265
NAME OF TOOL	SK. NO.
Drill Jig	9375
Plug Gage	9376
Pilot counterbore	9377
Gage	9378

FIG. 9. REFERENCE FILE CARD.

OPERATION SHEET.

Chemical Analysis.

3½% nickel steel.
C.—0.18 to 0.28 %.
Mn.—0.50 to 0.90%.
Ni.—3.00 to 3.75%.
P.—0.030 or less.
S.—0.35 or less.

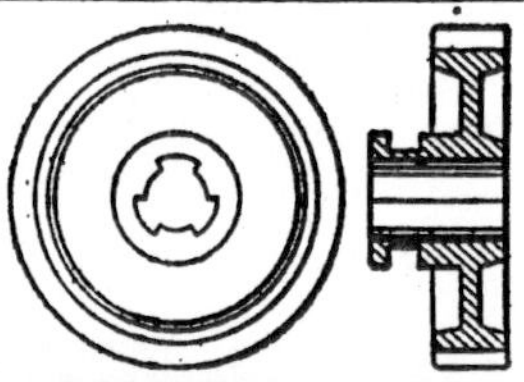

Draw No.—26,735.
Name.—Low Speed Gear
Date.—Sept. 30, 1910.
Superintendent.—O.K.
Gen. Foreman.—O.K.
Tool Designer.—O.K.

Oper. No.	Name of Operation.	Name of Machine Planned For.	Send to Dept.	Name of Tool.	Sketch No. of Tool.	Remarks.
1.	Anneal—put in furnace and heat to 1400° Leave in and cool down with furnace for about 10 hours.......		Hardening Dept.			
2.	Rough grind...........		Rough grinding.			
3.	Bore and ream "A," rough turn and face "E" and form "F"...	Potter & Johnston.		Standard tools. Form tools and bars.	 9,200	
4.	Face "H" and form "G."	Potter & Johnston.		Form tool and bar. Snap gage. Test plate.	9,201 9,202 9,203	
5.	Rough cut teeth "C"...	Gould & Eberhardt.		Collars. Backing plates.	9,204 9,205	
6.	Anneal—put in furnace and heat to 1400°. Leave in and cool down with furnace for about 10 hours.......		Hardening Dept.			
7.	Broach hole "A" to size.................	La Pointe broach...		Roughing broach. Finishing broach. Plug gage.	9,206 9,207 9,208	
8.	File off burrs left by broaching operation..		Bench.			
9.	Finish, turn and face "H" and "E." also groove "B" to size....	Engine lathe.		Taper arbor and bush. Formed lathe tool.	9,209 9,210	
10.	Inspect for gear cutter..		Inspection.			
11.	Finish cut teeth "C"...	Fellows gear shaper.		Test plate. Collars.	9,211 9,212	
12.	Round mill teeth "C"..	Tooth rounding.		Adaptors.	9,213	
13.	Bur and file teeth "C"..		Bench.			
14.	Inspect for hardening...		Inspection.			
15.	Pack harden, allow to cool in pot, reheat to 1375°-1425° F. and quench in oil. Inspect for hardness.........		Hardening Dept.			
16.	Wash in hot soda-water tank until all oil, etc.. is removed, then wire brush...............		Rough grinding.			
17.	Broach hole "A" to size if necessary..........	Hydraulic press.		Finishing broach. Plug gage. Limit gage for teeth.	9,214 9,215 9,216	
18.	Send to inspection department for final inspection............		Inspection Dept.	Plug gage. Test plate. Arbor. Scleroscope reading =	9,217 9,218 9,219	
19.	Send to gear department to be lapped.........		Gear Dept.	Lapping fixture.	9,220	

FIG. 10. OPERATION SHEET.

be distributed among the different departments without any delay.

When the tool is completed it is then tried out with the part and sent to the tool inspector to be checked up with the sketch and, after receiving his O. K., is sent to the tool crib or store, ready for use.

STANDARD PARTS FOR JIGS

A tool designer or maker should have a set of standard tables for his reference which insure the producing of a jig and its component parts all in proportion. It also reduces the cost of the jig and helps delivery of same. For example, let us consider a tool designer having a set of standard angles, channels, round plate and round stock, bricks, plates and square stock, as given in the accompanying tables, Nos. 1 to 10. The proportion and range of sizes have been compiled for jig making only and it is surprising how many jigs or fixtures can be built up from them.

As an example of the way these standard castings can be used refer to Fig. 1, which shows quite an elaborate drill jig. This is made up of practically two standard channels, a plate and standard bushings. For operating these bushings the standard wrenches can be used as shown. The large channel *A* has half of one side cut out and both the sides planed down to receive the plate *C*, which ties the sides of the channel together and also carries a drill bushing as shown.

A smaller channel *B* is planed to slide under the gibs *E E* and allows the bushings to be adjusted as desired with reference to those carried in *A*. Almost innumerable combinations can be made in this way and this also gives all examples of the perspective sketching in connection with this system. Tables 11 to 16 inclusive show standard sizes for handles and size bolts.

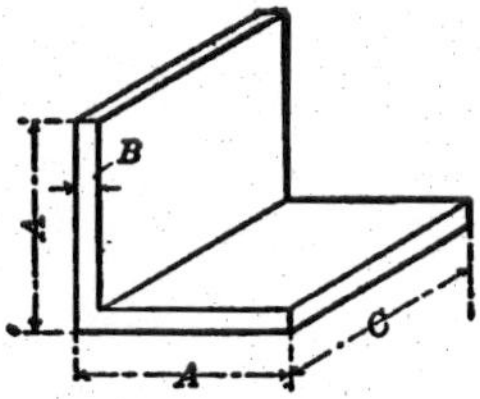

A	B	C
3″	⅞″	6″
4″	⅞″	6″
6″	1″	6″
6″	1¼″	12″
8″	1″	8″
8″	1¼″	12″
10″	1¼″	10″
12″	1¼″	12″
14¼″	1⅛″	7″
13″	1¼″	12″

TABLE 1. STANDARD ANGLE.

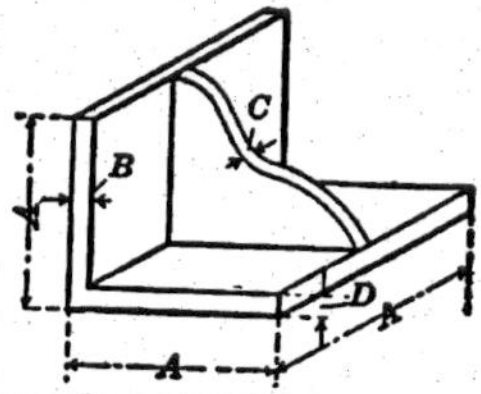

A	B	C	D
6″	1″	⅝″	1″
6″	1½″	⅞″	1″
8″	1″	⅝″	1″
12″	1¼″	⅞″	1″

TABLE 2. STANDARD ANGLE.

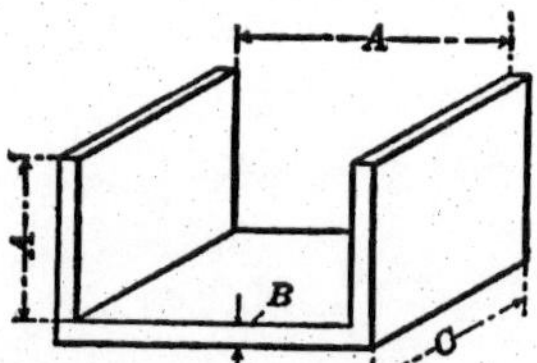

A	B	C
1¼″	¾″	6″
1¾″	⅞″	8″
2¼″	1″	10″
3″	⅞″	12″
4″	1″	14″
5″	1¼″	14″
6″	1¼″	20″
7″	1¼″	20″

TABLE 3. STANDARD CHANNEL.

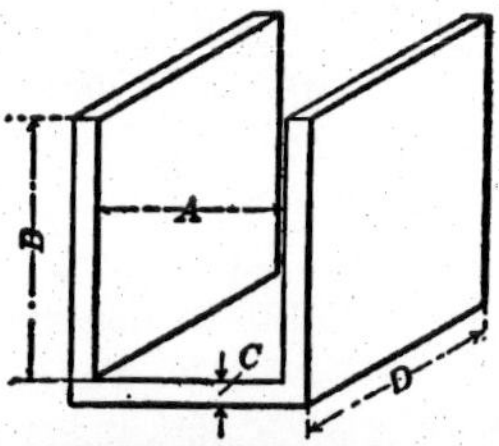

A	B	C	D
1¼″	3″	¾″	6″
1½″	4″	1⅜″	14″
1¾″	4″	⅞″	8″
2¼″	5″	1″	12″
2¾″	6″	1⅛″	12″
3½″	7″	1″	12″
4″	8″	1″	14″
5″	7″	1¼″	13″
6¼″	4″	1¼″	7″
6¾″	5″	1″	8″
7¾″	6″	1¼″	6″

TABLE 4. STANDARD CHANNEL.

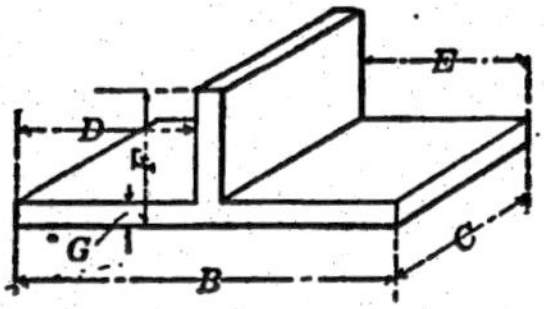

A	B	C	D	E	F	G
$2\frac{1}{8}''$	$10\frac{3}{8}''$	$5\frac{1}{8}''$	$2\frac{5}{16}''$	$5\frac{15}{16}''$	$6\frac{5}{8}''$	$1\frac{5}{16}''$
$2\frac{1}{4}''$	$14\frac{1}{2}''$	$6\frac{1}{8}''$	$6\frac{1}{4}''$	$6\frac{1}{4}''$	$7\frac{3}{4}''$	$1\frac{3}{8}''$

TABLE 5. STANDARD TEE.

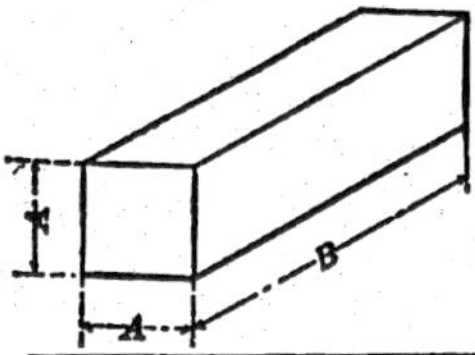

A	B	A	B
1″	6″	3″	12″
$1\frac{1}{2}''$	10″	4″	12″
2″	10″	5″	12″
$2\frac{1}{2}''$	12″		

TABLE 6. STANDARD SQUARE.

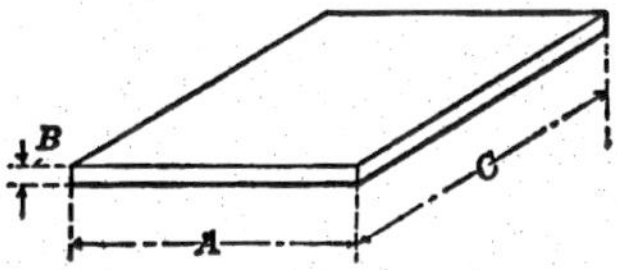

A	B	C
3″	1″	10″
3″	1″	22″
4″	1″	12″
4″	1″	24″
5″	$1\frac{1}{2}''$	12″
6″	1″	12″
6″	$1\frac{1}{4}''$	34″
7″	$1\frac{5}{8}''$	48″
8″	1″	12″
9″	$1\frac{1}{8}''$	14″
10″	$1\frac{1}{8}''$	24″
12″	$1\frac{1}{8}''$	24″
12″	$1\frac{3}{4}''$	12″
13″	$1\frac{1}{2}''$	28″
14″	1″	14″
16″	$1\frac{1}{4}''$	16″
18″	1″	18″
20″	$1\frac{1}{8}''$	20″

TABLE 7. STANDARD PLATE.

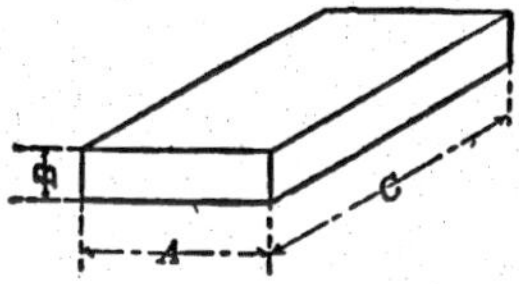

A	B	C
3″	$1\frac{1}{2}''$	6″
$3\frac{1}{2}''$	2″	7″
4″	$2\frac{1}{2}''$	8″
4″	2″	10″
$3\frac{1}{4}''$	$2\frac{1}{4}''$	$15\frac{1}{4}''$
$4\frac{1}{2}''$	$2\frac{1}{2}''$	9″
7″	$2\frac{1}{2}''$	12″
9″	5″	18″

TABLE 8. STANDARD BRICK.

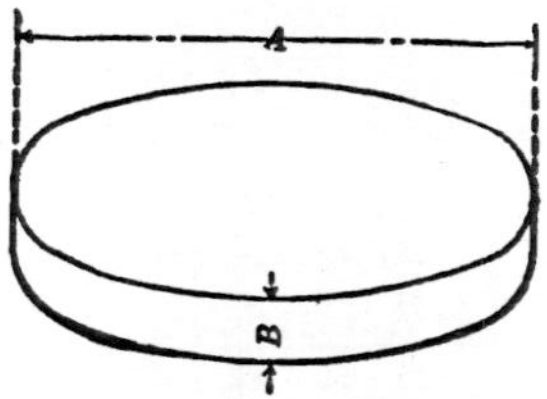

A	B
8	1
10	1½
12	2
14½	1½
16	2¼
22	1

TABLE 9. STANDARD ROUND.

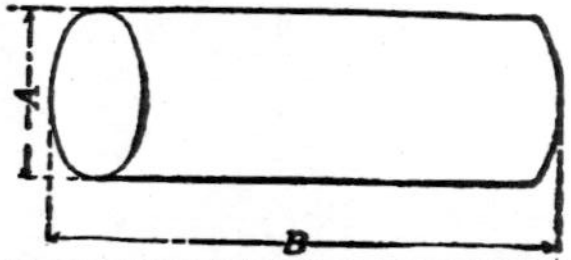

A	B
¾	10
1	4
1½	8
2	6
2½	10
3	8
4	10
5	10
6	12

TABLE 10. STANDARD BAR.

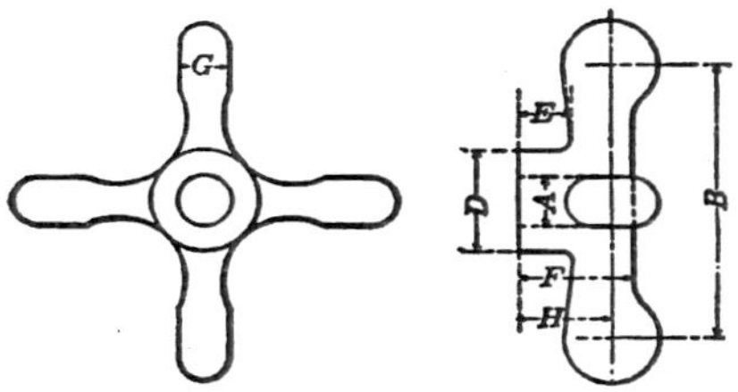

A	B	C	D	E	F	G	H
1/4″	1 5/16″	5/32″	1/2″	5/16″	9/16″	1/4″	7/16″
5/16″	1 11/16″	7/32″	9/16″	5/16″	9/16″	5/16″	1/2″
3/8″	2″	5/16″	11/16″	3/8″	3/4″	3/8″	5/8″
7/16″	2 5/16″	11/32″	3/4″	5/16″	3/4″	3/8″	9/16″
1/2″	2 7/16″	3/8″	7/8″	3/8″	7/8″	3/8″	5/8″

TABLE 11.
MALLEABLE IRON CROSS JIG KNOBS.

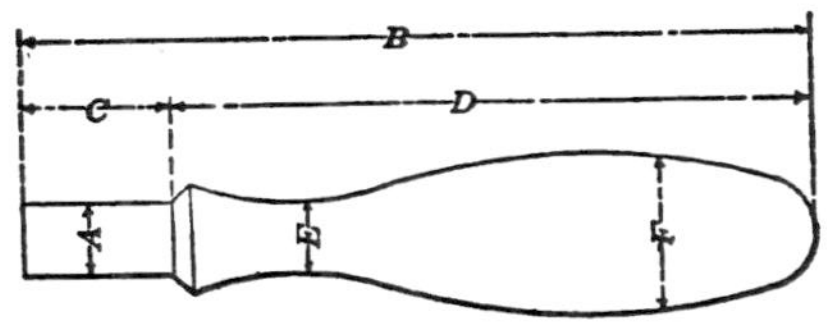

A		B	C	D	E	F
Rough Diam.	Finish Diam.					
7/16″	3/8″	2 9/16″	19/32″	1 31/32″	3/8″	7/8″
1/2″	7/16″	3 1/2″	3/4″	2 3/4″	15/32″	1″
1/2″	7/16″	3 3/4″	3/4″	3″	15/32″	1 1/8″
1/2″	7/16″	4 1/2″	7/8″	3 5/8″	17/32″	1 1/4″
9/16″	1/2″	5″	1″	4″	17/32″	1 5/16″
11/16″	5/8″	5 3/4″	1″	4 3/4″	21/32″	1 1/2″

TABLE 12. JIG HANDLES—STRAIGHT.

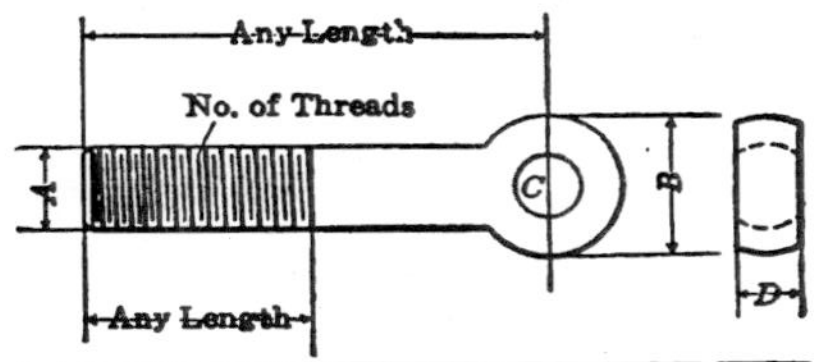

A	No. of Threads.	Form of Threads.	B	C	D
1/4″	20	U.U.S.	7/16″	3/16″	1/4″
5/16″	18	U.U.S.	9/16″	1/4″	5/16″
3/8″	16	U.U.S.	5/8″	5/16″	3/8″
7/16″	14	U.U.S.	3/4″	3/8″	7/16″
1/2″	13	U.U.S.	7/8″	7/16″	1/2″
9/16″	12	U.U.S.	1″	7/16″	9/16″
5/8″	11	U.U.S.	1 1/8″	1/2″	5/8″
3/4″	10	U.U.S.	1 1/4″	9/16″	3/4″
7/8″	9	U.U.S.	1 1/2″	5/8″	7/8″
1″	8	U.U.S.	1 3/4″	3/4″	1″

TABLE 13. EYE BOLT ENDS.

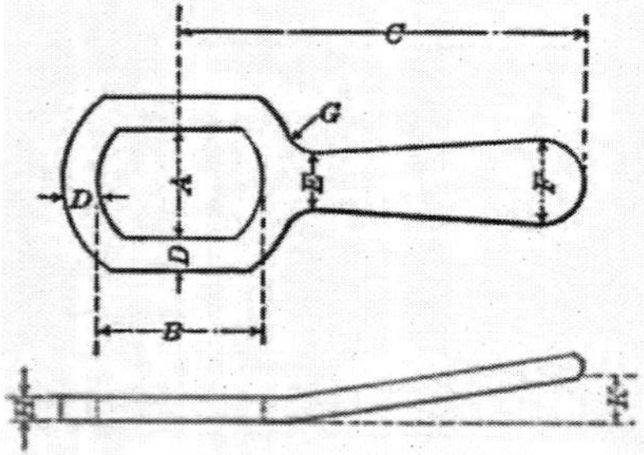

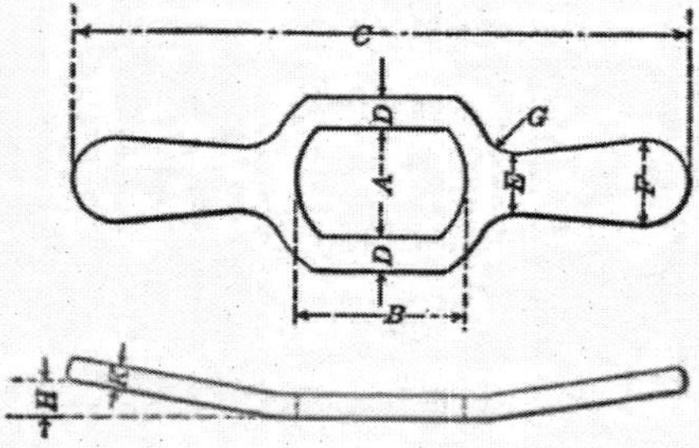

Single Handle.									Double Handle.								
A	B	C	D	E	F	G	H	K	A	B	C	D	E	F	G	H	K
7/8"	1 3/8"	3 1/2"	5/16"	1/2"	3/4"	1/2"	1/4"	1/2"	2"	2 1/2"	9"	7/16"	7/8"	1"	1/2"	1/2"	3/8"
1"	1 1/2"	3 1/2"	5/16"	1/2"	3/4"	1/2"	1/4"	1/2"	2 1/4"	2 3/4"	9"	7/16"	7/8"	1"	1/2"	1/2"	3/8"
1 1/4"	1 3/4"	4"	5/16"	5/8"	7/8"	1/2"	1/4"	1/2"	2 3/4"	3 1/4"	10"	1/2"	7/8"	1"	1/2"	1/2"	3/8"
1 3/8"	2 1/8"	4 1/2"	3/8"	3/4"	1"	1/2"	1/4"	1/2"	3"	3 1/2"	10"	5/8"	7/8"	1"	1/3"	1/2"	3/8"

TABLE 14. SINGLE AND DOUBLED HANDLED JIG WRENCHES.

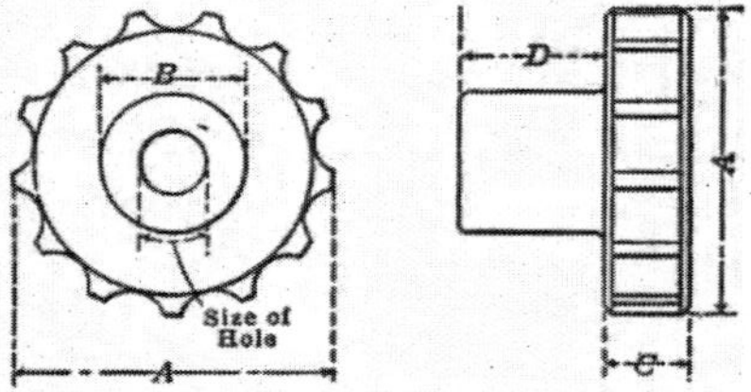

Size of Hole.	A	B	C	D
1/4"	1 1/2"	3/4"	3/8"	1/2"
5/16"	1 1/2"	3/4"	3/8"	1/2"
3/8"	1 1/2"	3/4"	3/8"	1/2"
7/16"	1 1/2"	3/4"	3/8"	1/2"
1/2"	2"	1 1/16"	1/2"	1/2"
9/16"	2"	1 1/16"	1/2"	5/8"
5/8"	2"	1 1/16"	1/2"	5/8"
3/4"	2 1/2"	1 1/4"	5/8"	5/8"
7/8"	2 1/2"	1 1/4"	5/8"	5/8"

TABLE 15. MALLEABLE IRON JIG KNOBS.

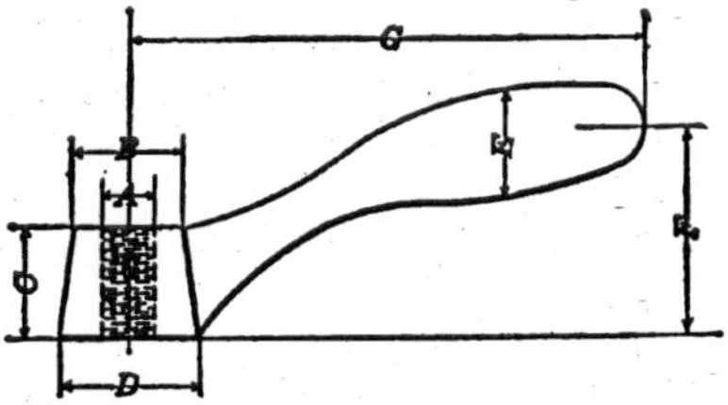

A	B	C	B	E	F	G
1/4″–24	1/2″	1/2″	5/8″	1/2″	15/16″	2 3/8″
5/16″–18	1/2″	1/2″	5/8″	1/2″	15/16″	2 3/8″
3/8″–16	1/2″	1/2″	5/8″	1/2″	15/16″	2 3/8″
7/16″–14	3/4″	3/4″	1″	11/16″	1 1/16″	3 1/4″
1/2″–13	3/4″	3/4″	1″	11/16″	1 1/16″	3 1/4″
9/16″–12	1″	7/8″	1 1/4″	3/4″	1 1/4″	4 1/2″
5/8″–11	1″	7/8″	1 1/4″	3/4″	1 1/4″	4 1/2″
3/4″–10	1″	7/8″	1 1/4″	3/4″	1 1/4″	4 1/2″

TABLE 16. JIG HANDLES—BENT.

CHAPTER II

THE DIFFERENT KINDS OF JIGS

The jig shown in Fig. 11 is generally classed as a temporary jig. It is located on a casting by means of two or more locating pins and is held by a clamp, while the holes are being drilled.

For small jigs the plate *A* is often made of steel, and instead of having bushings this plate is drilled out to suit the drill to be used, and then it is case hardened, after which the holes are lapped and the locating pins driven in place. They are more frequently made of cast iron, in which case the locating pins are hardened and the standard form of bushing used.

Fig. 12 is an open-face jig, and generally used for flat work. The part enters the front of the jig and is forced against the locating points *B* by the screw *C* and held down by the screw *A*. For removing the part, a hole *D* can be placed in the back of the jig, through which the operator can enter a screwdriver or round piece of stock, and quickly work the part out.

Fig. 13 shows what is generally called a pan jig, and is used for drilling small plates and it also demonstrates the use of handles on jigs. The handle *C* acts as a quick means for moving the jig about and holds the jig while the screw *A* is being tightened. The plate *B*, which contains the guide bushings, acts as a clamp and is removable when a screw *A* is turned half way around.

Fig. 14 may be called a hex jig and is used for drilling hex head bolts, nuts, collars, etc. The illustration shows a jig designed for drilling wrench holes around a collar. It is made of standard hexagon steel. The part is held in position by the screw *A*, and is forced out by means of the plunger *B*.

Fig. 15 is a simple design for jig for drilling work such as pins. It can be made of round steel and flattened on the bottom and provided with a knockout plunger *B*, similar to Fig. 14. The part is held in position by means of the screw *A*.

Fig. 16 is similar to Fig. 15, and is often used for drilling pins, but more frequently for bar or tubing. It is made of square steel stock. The part is held in position by means of the screw *A* and is located against a stop pin *B*. The advantage this has over Fig. 15 is that one hole can be drilled at right angles with another. In this design it is well to remember to relieve the jig opposite to drill

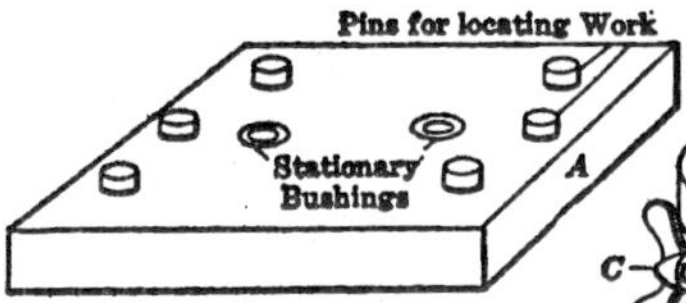

FIG. 11.

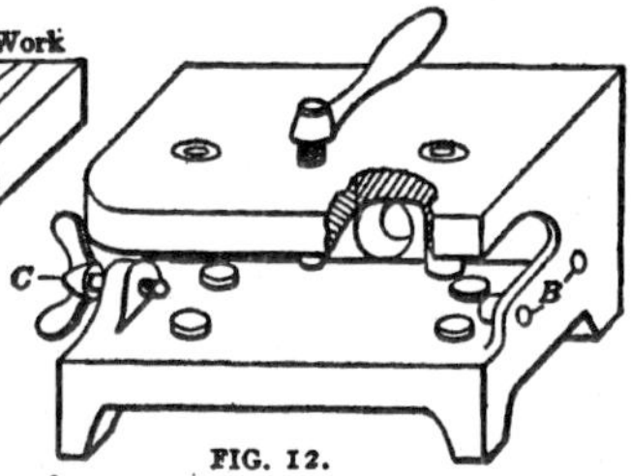

FIG. 12.

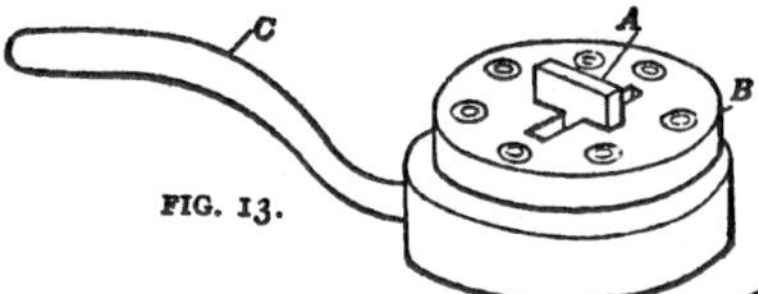

FIG. 13.

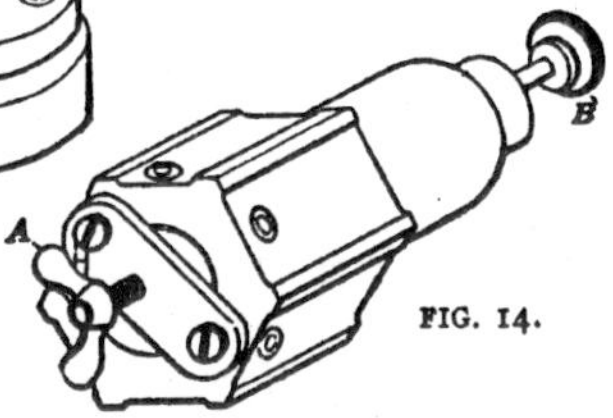

FIG. 14.

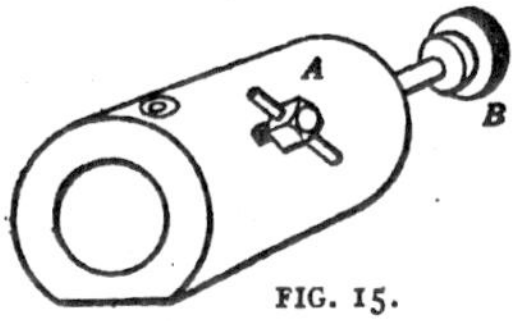

FIG. 15.

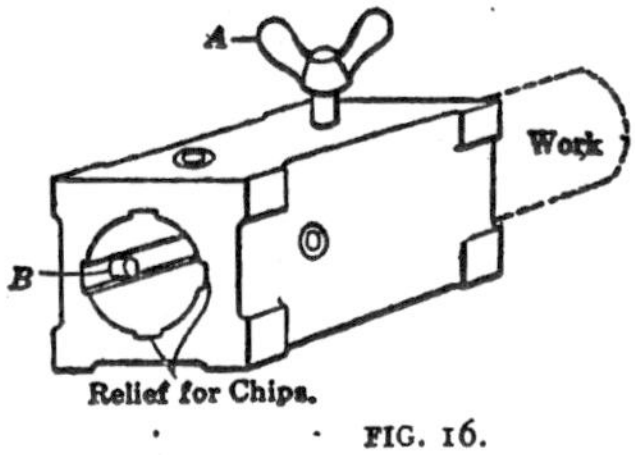

FIG. 16.

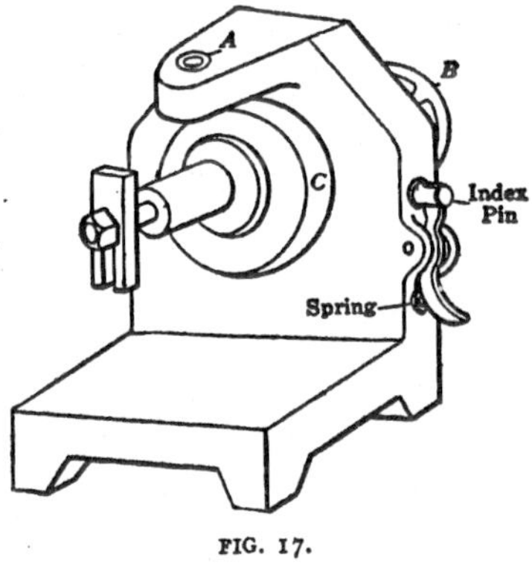

FIG. 17.

FIGS. 11 TO 17.—A FEW SIMPLE DRILLING JIGS.

bushing, to clear the bur raised by the drill and allow the piece to be freely drawn from the jig.

Fig. 17 shows a rotary-index jig. The part in this case may have any number of holes around the outside diameter and the jig is indexed to correspond. All holes can be drilled through the stationary bushing *A*. The part is fastened to the plate *C* and revolved, the indexing pin in the side does the spacing. A good-sized hand wheel *B* should be provided for turning plate *C*.

Fig. 18 may be called a double-acting rotary jig. The piece to be drilled is clamped in *A*, which revolves on *B*, as shown by

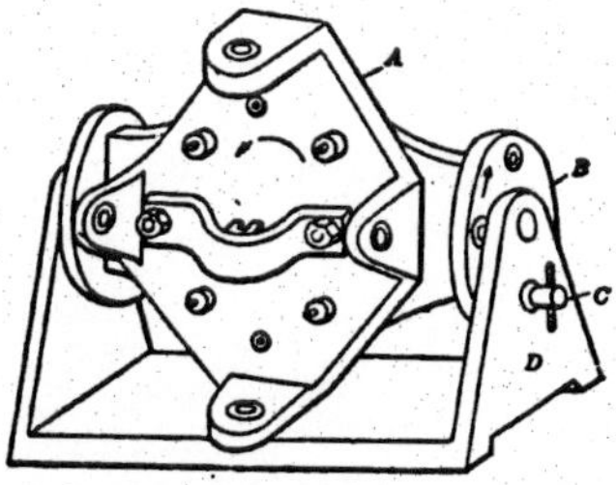

FIG. 18.

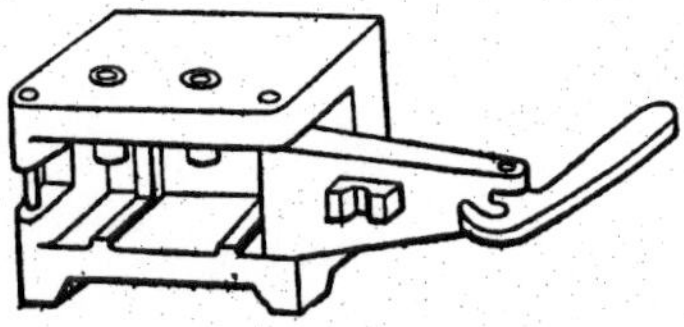

FIG. 20.

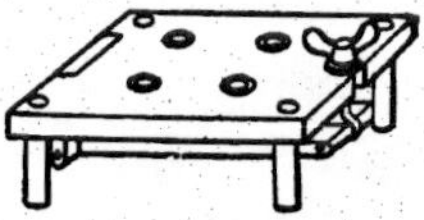

FIG. 19.

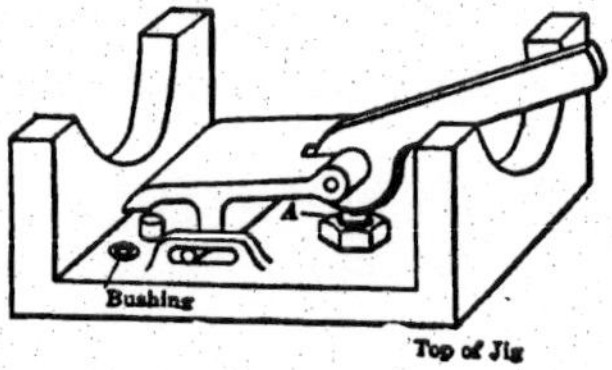

FIG. 21.

FIGS. 18 TO 21.—INDEXING AND LATCH JIGS.

arrow. *B* revolves in the cradle *D* and is held in position by the index pin *C*. This style of jig is very good for drilling large work and will be found a great help to the operator. It is very necessary that the jig be well balanced.

Fig. 19 is a table jig. The design shows a built-up jig, which is generally done in a case of small work, but for large jigs of this type, it is well to have the legs cast integral with the plate. In the design shown, the legs are made of cold-drawn steel, which, when hardened, are driven in place. Some designers screw them in, but this is not necessary, and is expensive. When the legs are in place they are surface ground square with the plate of the jig. Legs should always be made as short as possible and their face as large as is consistent with the design. A great mistake is often made in making them smaller than the slot in the drill-press table, and also in not properly ribbing a cast leg, or of making it too light. The work should be in all cases well clamped to the plate as the thrust of the drill is, in this case, directly against the work, and has a tendency to spring it.

Fig. 20 shows an open-end jig, but is called a hook cam-lever jig, simply to demonstrate the use of this style of cam. It is unnecessary to describe the advantages of this style of clamping, as they are too well known. Many tool designers do not use this style of clamping often enough, because of fear of it working loose. When they do, it is not the fault of the cam lever, but rather the design.

Fig. 21 shows another design of cam-lever jig, and for convenience it is called a standard cam-lever jig, as it is more frequently used than the one in Fig. 20. Its advantages are the same as mentioned in Fig. 20, but as a word of caution, it may be well to mention that the cam lever should not be used where there is apt to be any great variation in the part to be clamped as the cam eccentricity is not usually over 1/16 of an inch. The cam lever can, however, be used as an adjusting screw *A* is placed under the cam part of the lever, as shown.

Fig. 22 is called a wedge jig to show the use of the wedge for holding the work. It is used as often as the cam lever. Wedges, like cam levers, have only a limited amount of movement and taper, and therefore can be used only where the amount of variation in the piece is within this limit.

Fig. 23 is an eccentric-pin jig and shows how an eccentric pin is used for holding the piece. The pin in this case is shown loose, but in some instances it is held in position by means of a washer and nut. This, however, is made to suit the conditions of the jig. By making the pin loose, the work can be placed in or removed from the jig quickly.

Fig. 24 is a simple form of jig, showing the use of a sliding lid. In some instances these lids contain the drill bushings as well as clamping screw, in which case it is located by means of an index pin. The illustration shows the lid *A* used only to carry the clamping screw *B*, which holds the work to be drilled. When removing the work from the jig, the screw is released and the lid is slid out of the jig, the pin *C* acting as a stop for the lid.

Fig. 25 shows two slip bushings *A* and *B*, which hold the work as well as piloting the drill. Frequently a spring is placed under the head of these bushings so that when the pressure of the bushings is released, they will automatically rise and allow the part to be quickly removed.

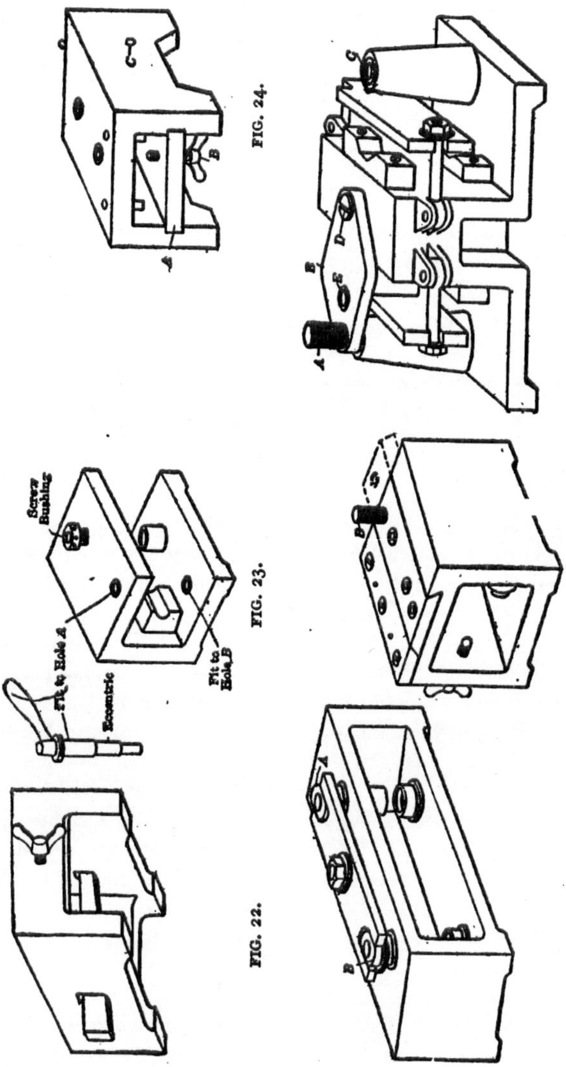

FIG. 22. FIG. 23. FIG. 24. FIG. 25. FIG. 26. FIG. 27.

FIGS. 22 TO 27.—WEDGE, ECCENTRIC PIN AND SLIDING WEDGE JIGS.

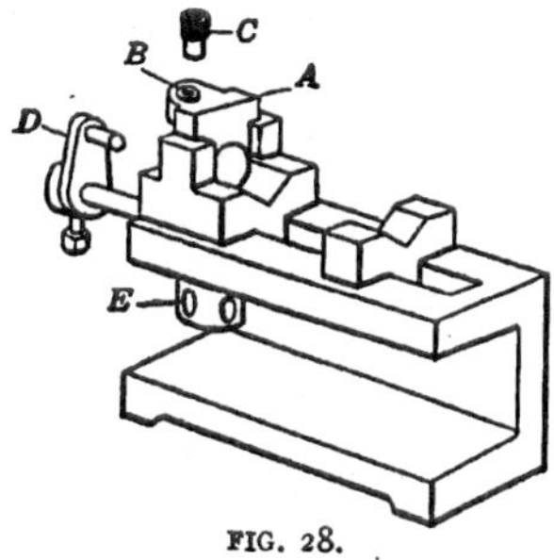

FIG. 28.

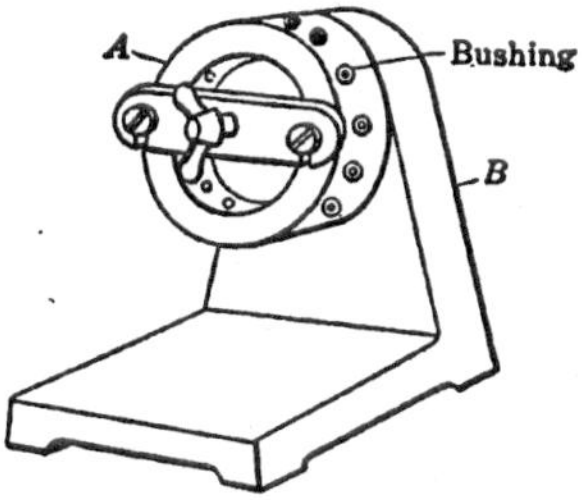

FIG. 29.

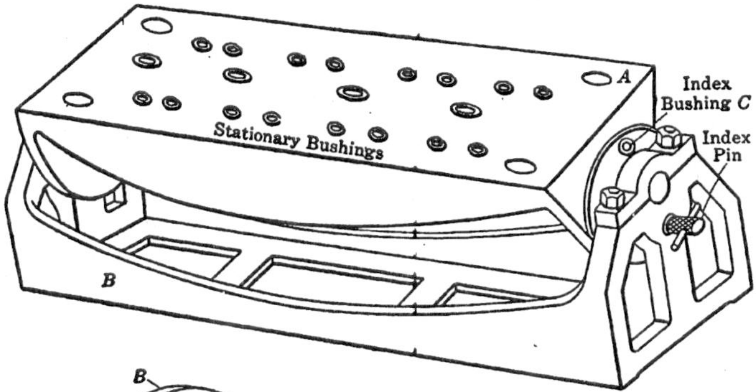

FIG. 30.

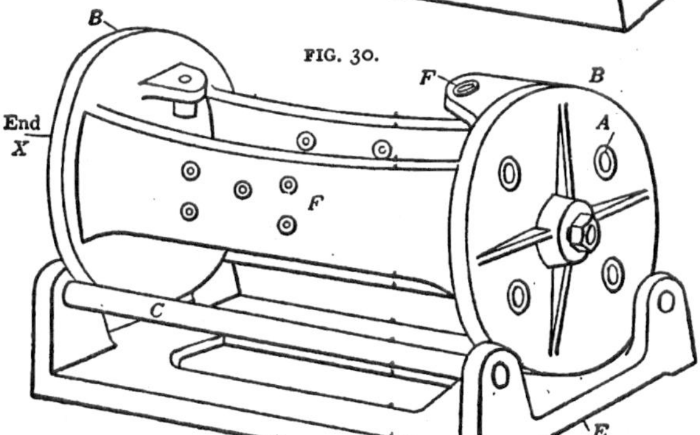

FIG. 31.

FIGS. 28 TO 31.—BAR STOCK, ROTARY AND CRADLE JIGS.

Fig. 26 is called an index-slide jig, and is used for drilling a series of holes close together. The slide *A* contains the drill bushings and should be a sliding fit in the gibs. It is indexed by the pin *B*, which is removable.

Fig. 27 shows a reciprocating jig. While the work is being drilled on the left-hand side, the operator is taking out and inserting another piece on the right-hand side. The plate *B*, holding the drill bushing *E*, swings on the screw *D* and is located by means of the index pin *A* and bushing *C*.

Fig. 28 shows a jig for drilling holes in round bar stock. The capacity of this style of jig is equal to the size of the hole in the clamping slide *A*, which carries a stationary bushing *B*. In this bushing is inserted a slip bushing *C*, which is the size of the drill corresponding to the size of the hole to be drilled in the bar. A number of these slip bushings can be made up with various-sized holes. The work rests in the Vs, and is located by the adjustable stop *D*. The part is held in the Vs by the clamping slide *A*, a nut *E* operates this slide. The jig shown is an improvement over jigs of this character.

Fig. 29 is another design of rotary jig of the larger class. The work is bolted to the revolving frame or body *A*, which is supported by a cradle *B*. The indexing of it is done in the usual manner by means of the bushings *C* and index pin *D*; like all rotary jigs it is essential that they be in balance, otherwise the operator will have difficulty in turning it over.

Fig. 30 shows a small rotary jig. The drum *A*, in which the work to be drilled is placed, revolves in either direction on the stud, which is part of the base *B*. It will be noted in this case that the drum is not indexed, the operator, therefore has to hold it while the drilling is being done.

Fig. 31 shows the design of an upright as well as revolving jig and its cradle. For drilling the holes *F* around the outer surface, it is placed in the cradle *E*, which has two rollers attached to it; the ends *B B* are turned a true circle to allow it to be freely turned around by hand. For drilling the holes *A*, it is taken out of the cradle and set on the end.

Fig. 32 is still another style of rotary jig and one that may be set upon the table of the drill press, as shown in the illustration. The revolving member or body *A* hangs over the side of the table. The angle *B*, which supports the body *A*, is fastened to the drill-

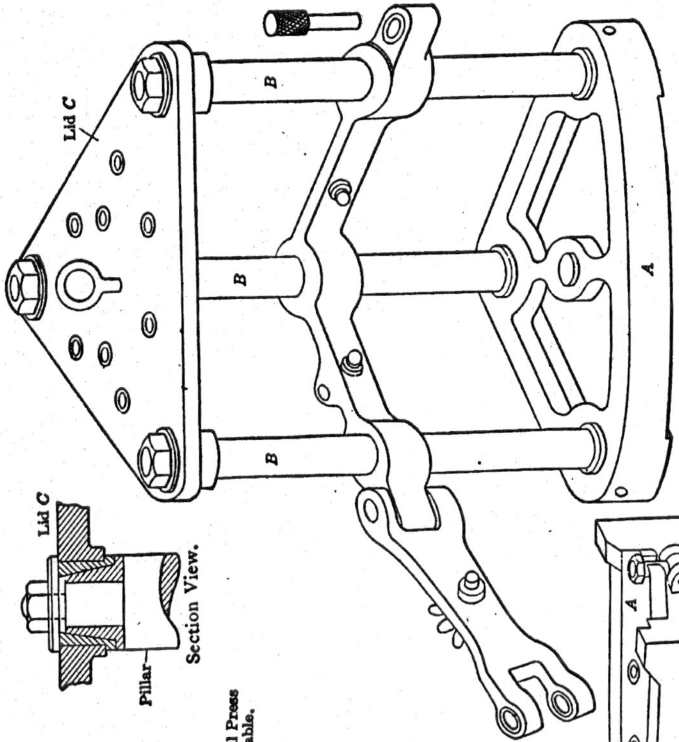

FIG. 33.

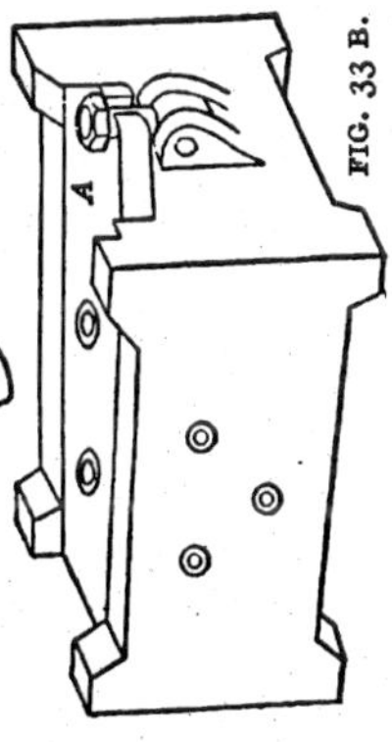

FIG. 33 B.

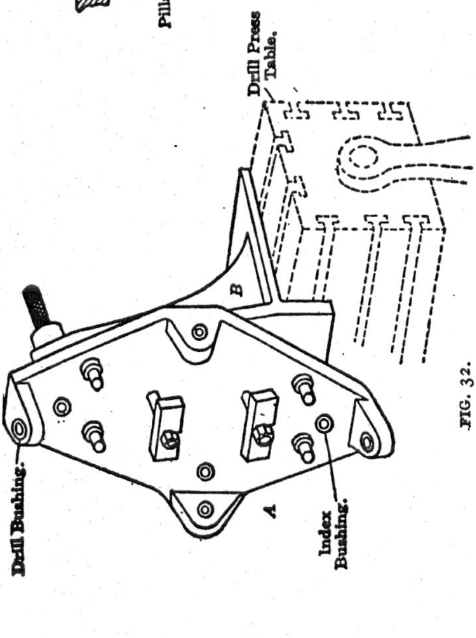

FIG. 32.

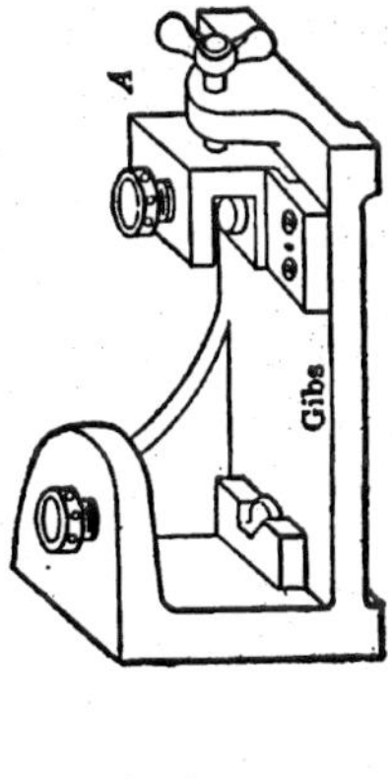

FIG. 33 A.

press table in the usual manner. This style of jig is generally made up for large pieces in which small holes are to be drilled around its outer surface. The indexing is arranged the same as shown in Fig. 29.

Fig. 33, called a pillar jig, is an excellent way of making a jig for very large work. The pillars *B B B* are made of steel, driven in the base *A* and pinned. The ends of these pillars have a hardened collar which is slightly tapered, and *C* shows a corresponding bushing, as shown in the section view. With this arrangement the plate can be readily placed on or removed from the pillars. The weight of a large jig of this description can be reduced to a minimum and is approximately one-third lighter than an all-casting jig.

Fig. 33A, which is called a double-acting jig, demonstrates the use of the screw bushings and a stationary and movable V. This style of jig is made for drilling holes such as in levers, which generally have a limit of plus or minus 1/16 of an inch. The part is centered by two Vs; one is stationary and *A* carries the movable one as well as the screw bushing. When the work is centered it is held down by the screw bushings.

Fig. 33B shows a box type of jig. The lid *A* is of the swinging style, lids of different kinds being used in this connection. This style of jig is generally made when the part has a number of holes to be drilled on all sides.

CHAPTER III

LOCATING SCHEMES FOR DRILL JIGS

Many methods are employed to center or locate the work, there being no special design which is best for all cases. The plans shown herewith give a variety of ways in which this can be done and allow the selection of the one the designer considers best for any particular case.

One of the most useful means employed in jig construction for locating or centering the part to be drilled is the V, its construction varying to suit conditions. It is usually operated by means of a screw with a double or triple thread.

Fig. 34 shows a simple construction that is generally used on very small jigs. The V is held down by means of the screw *A* and is also guided by this screw, which is flattened on both sides to fit the slot in V. As shown, the end of the screw also acts as a guider.

Fig. 35 is an improved design, being guided by the two gibs *A* and *B*. The V is held down in the same manner as in Fig. 34 by means of the screw *C*. This design is also used in small-jig construction.

A CHEAPER FORM

Fig. 36 is practically the same as Fig. 35 except it is a little cheaper to make. It is held down the same way as Fig. 34 but is guided by a tongue instead of the gibs. There is a slot cut in the body of the jig in which this tongue slides.

Fig. 37 is really the standard form of design for constructing these Vs as it is strong and accurate. The gibs *A A* should be dowel pinned after being located and should be made of steel. In some instances they are hardened and ground but generally left soft and the V hardened.

Fig. 38 is similar to Fig. 37 except in the design of the gibs, which are dovetailed. This style is more often used in large drill jigs and is, of course, more expensive than any of the others shown, but it has advantages over the others in the way of accuracy and wearing qualities.

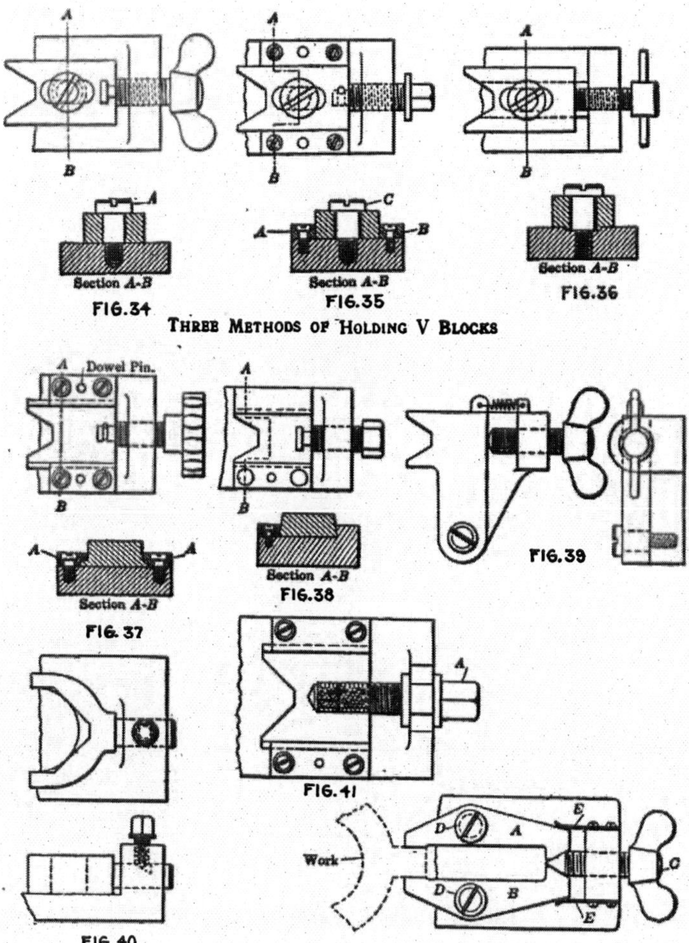

FIGS. 34 TO 42.—SEVERAL LOCATING SCHEMES.

Fig. 39 is another form of V which is often used when the design of the jig calls for a limited amount of room, being particularly useful in such cases.

USED ON LARGE JIGS

Fig. 40 is a satisfactory way of making a V for locating large bosses, etc. In this design the V has only a limited amount of movement. It is not very popular and is used mostly on large jigs.

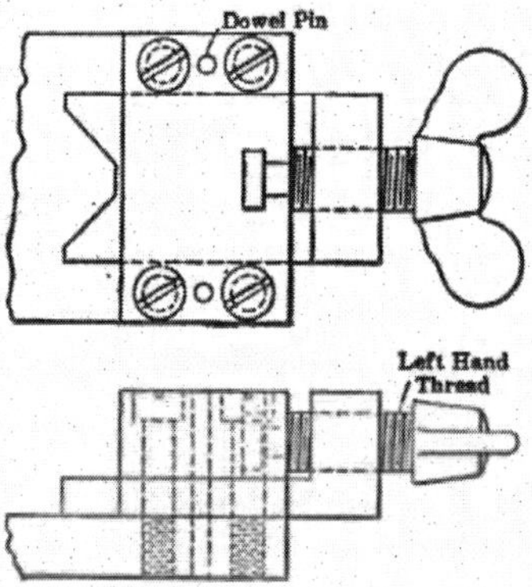

FIG. 43.

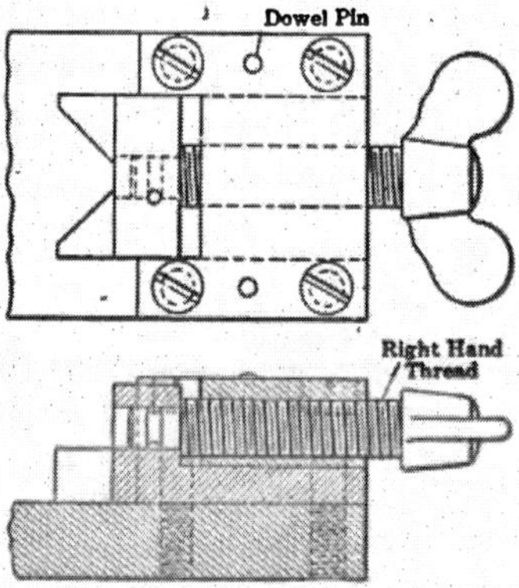

FIG. 44.

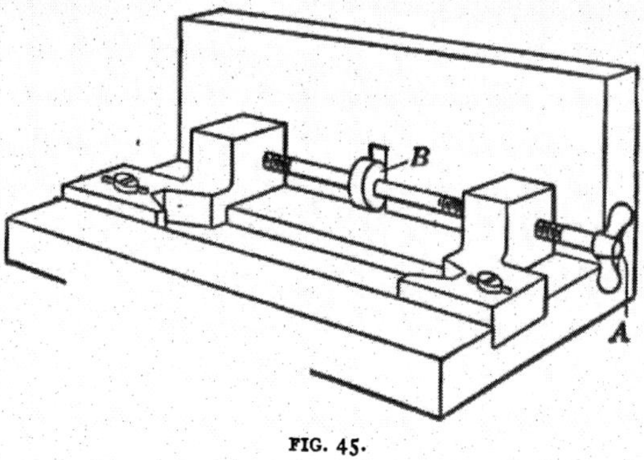

FIG. 45.

FIGS. 43 TO 45.—MORE LOCATING SCHEMES.

Fig. 41 is similar to Fig. 37 and is only shown to give the designer or toolmaker another method of operating a V. The screw *A* is threaded into the block and held by the two collars shown.

Fig. 42 consists of two fingers *A* and *B* which are operated by means of a screw *C*. These fingers swing on the screws *D* and are brought back to their original position by means of springs *E E*. This is another design which can be used to advantage when there is only a limited amount of room in the jig.

Figs. 43 and 44 show two styles of locating Vs which are held down in similar ways but are operated somewhat differently. In the first case the back end of the V block acts as the nut, the screw not having any end movement. In the second the screw works through the cross bar and is held in the front end of the V. The gib in both cases acts as a guide for the V and support for the screw.

In Fig. 45 are shown two Vs operated by a screw *A* which has a right- and left-hand thread. The flange *B* in the center of this screw works in a slot which holds it in a central position. The two Vs are guided in a slot provided for them in the jig and are held down in the same manner as in Fig. 34.

It will be observed that each of the forms shown, as representative locating devices, possess some advantages for special purposes, and each will be found useful in certain places.

CHAPTER IV

STANDARD BUSHINGS FOR DRILL JIGS

It is obvious that economy in the tool room demands the selection of standard bushings so that they can be made at less expense and even in advance if necessary. But, unless a tool designer has had considerable experience in the use of bushings of various kinds, he is often at a loss to know what size bushings to select for each case.

The accompanying illustrations show about all the forms of bushings that experience has shown to be necessary in drilling jig work as well as the proportions which have proved satisfactory. These require little explanation, the illustrations and dimensions containing about all necessary information.

Fig. 46 shows the standard design of stationary jig bushings that are universally used in jig construction. Table No. 17 gives dimensions.

The length of bushings should be equal to twice the diameter of the drill. Small bushings should stand away from the work about 1/16 in. and the larger sizes, 1/4 in.

All stationary bushings should be slightly chamfered on one end; just enough to prevent shearing the hole when they are being driven in place.

The best practice in grinding a stationary bushing is to grind the end *A* until it is a sliding fit in the hole into which it is to be pressed. The length *C* of this end varies but is generally made 1/16 to 1/8 in. long. The diameter *B* of the bushing is ground straight with an allowance of 0.002 or 0.003 in. for driving.

The standard shouldered stationary jig bushing is generally used when the wall of the jig is thin and the bushing must project. It is also used to take up the thrust of a stop collar on a drill, facing tool or counterbore. The dimensions for bushings shown in Fig. 47 in Table No. 17 are given. The bushings should be ground straight and one-half to one-thousandth larger than the size of the hole into which they are to be pressed.

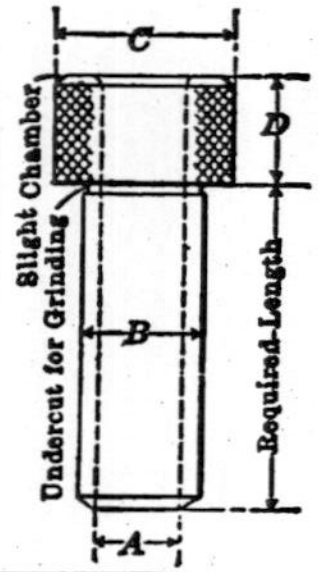

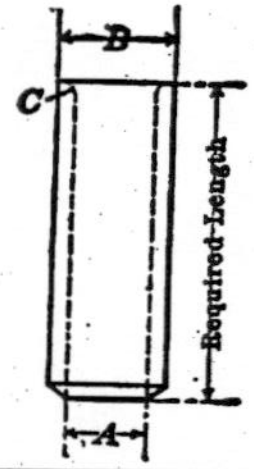

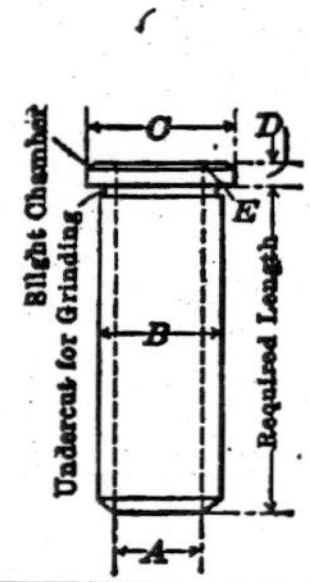

SLIP BUSHINGS.

A	B	C	D
$\frac{1}{4}$	$\frac{3}{8}$	$\frac{5}{8}$	$\frac{3}{8}$
$\frac{5}{16}$	$\frac{1}{2}$	$\frac{13}{16}$	$\frac{1}{2}$
$\frac{3}{8}$	$\frac{9}{16}$	$\frac{7}{8}$	$\frac{9}{16}$
$\frac{7}{16}$	$\frac{5}{8}$	$\frac{15}{16}$	$\frac{9}{16}$
$\frac{1}{2}$	$\frac{3}{4}$	$1\frac{1}{8}$	$\frac{9}{16}$
$\frac{9}{16}$	$\frac{13}{16}$	$1\frac{3}{16}$	$\frac{5}{8}$
$\frac{5}{8}$	$\frac{7}{8}$	$1\frac{5}{16}$	$\frac{5}{8}$
$\frac{11}{16}$	$\frac{15}{16}$	$1\frac{7}{16}$	$\frac{5}{8}$
$\frac{3}{4}$	$1\frac{1}{16}$	$1\frac{7}{16}$	$\frac{3}{4}$
$\frac{13}{16}$	$1\frac{1}{8}$	$1\frac{9}{16}$	$\frac{3}{4}$
$\frac{7}{8}$	$1\frac{1}{4}$	$1\frac{11}{16}$	$\frac{3}{4}$
$\frac{15}{16}$	$1\frac{5}{16}$	$1\frac{3}{4}$	$\frac{7}{8}$
1	$1\frac{3}{8}$	$1\frac{13}{16}$	$\frac{7}{8}$

STATIONARY BUSHINGS.

A	B	C
$\frac{1}{4}$	$\frac{3}{8}$	$\frac{1}{32}$
$\frac{5}{16}$	$\frac{1}{2}$	$\frac{1}{32}$
$\frac{3}{8}$	$\frac{9}{16}$	$\frac{1}{32}$
$\frac{7}{16}$	$\frac{5}{8}$	$\frac{1}{32}$
$\frac{1}{2}$	$\frac{3}{4}$	$\frac{1}{16}$
$\frac{9}{16}$	$\frac{13}{16}$	$\frac{1}{16}$
$\frac{5}{8}$	$\frac{7}{8}$	$\frac{1}{16}$
$\frac{11}{16}$	$\frac{15}{16}$	$\frac{1}{16}$
$\frac{3}{4}$	$1\frac{1}{16}$	$\frac{1}{16}$
$\frac{13}{16}$	$1\frac{1}{8}$	$\frac{3}{32}$
$\frac{7}{8}$	$1\frac{1}{4}$	$\frac{3}{32}$
$\frac{15}{16}$	$1\frac{5}{16}$	$\frac{3}{32}$
1	$1\frac{3}{8}$	$\frac{3}{32}$

SHOULDER BUSHINGS.

A	B	C	D	E
$\frac{1}{4}$	$\frac{3}{8}$	$\frac{1}{2}$	$\frac{3}{32}$	$\frac{1}{32}$
$\frac{5}{16}$	$\frac{1}{2}$	$\frac{5}{8}$	$\frac{3}{32}$	$\frac{1}{32}$
$\frac{3}{8}$	$\frac{9}{16}$	$\frac{11}{16}$	$\frac{3}{32}$	$\frac{1}{32}$
$\frac{7}{16}$	$\frac{5}{8}$	$\frac{3}{4}$	$\frac{1}{8}$	$\frac{1}{32}$
$\frac{1}{2}$	$\frac{3}{4}$	$\frac{7}{8}$	$\frac{1}{8}$	$\frac{1}{16}$
$\frac{9}{16}$	$\frac{13}{16}$	1	$\frac{1}{8}$	$\frac{1}{16}$
$\frac{5}{8}$	$\frac{7}{8}$	$1\frac{1}{8}$	$\frac{3}{16}$	$\frac{1}{16}$
$\frac{11}{16}$	$\frac{15}{16}$	$1\frac{3}{16}$	$\frac{3}{16}$	$\frac{1}{16}$
$\frac{3}{4}$	$1\frac{1}{16}$	$1\frac{5}{16}$	$\frac{3}{16}$	$\frac{1}{16}$
$\frac{13}{16}$	$1\frac{1}{8}$	$1\frac{3}{8}$	$\frac{1}{4}$	$\frac{3}{32}$
$\frac{7}{8}$	$1\frac{1}{4}$	$1\frac{9}{16}$	$\frac{1}{4}$	$\frac{3}{32}$
$\frac{15}{16}$	$1\frac{5}{16}$	$1\frac{11}{16}$	$\frac{1}{4}$	$\frac{3}{32}$
1	$1\frac{3}{8}$	$1\frac{3}{4}$	$\frac{5}{16}$	$\frac{3}{32}$

TABLE 17. THREE KINDS OF PLAIN BUSHINGS.

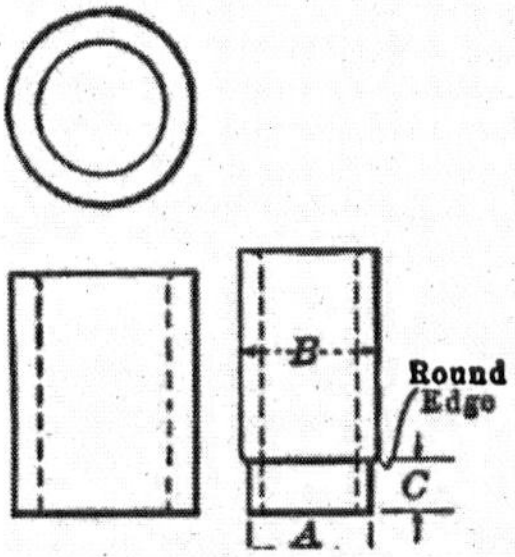

FIG. 46.

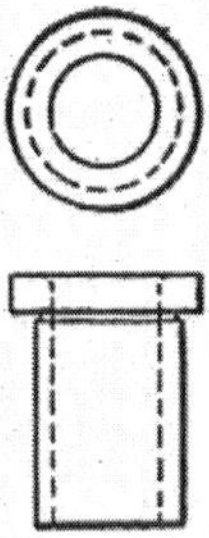

FIG. 47.

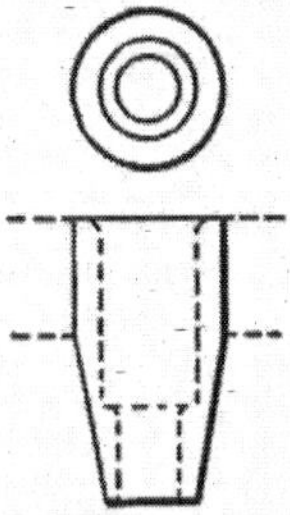

FIG. 48.

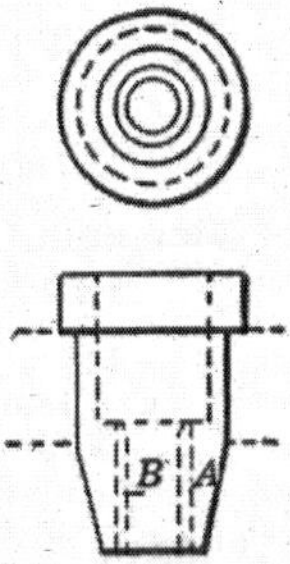

FIG 49.

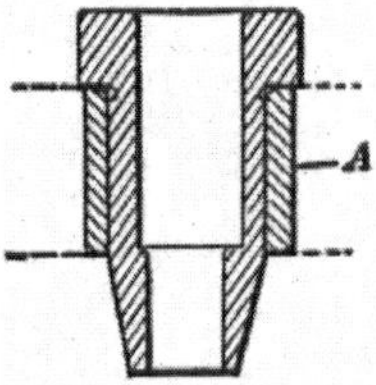

FIG. 50.

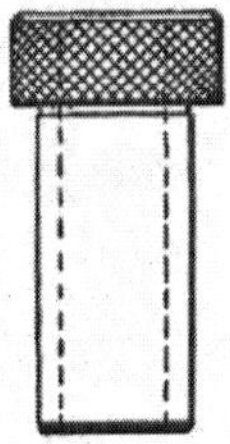

FIG. 51.

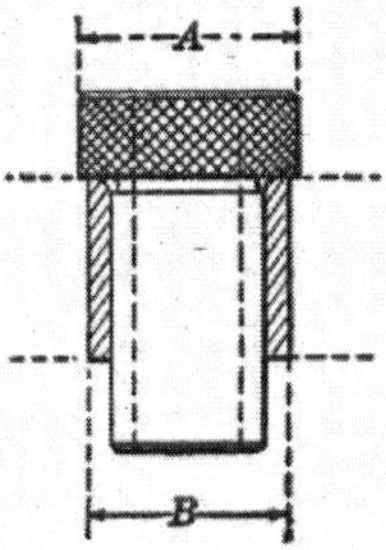

FIG. 52.

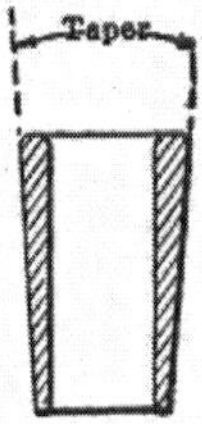

FIG. 53.

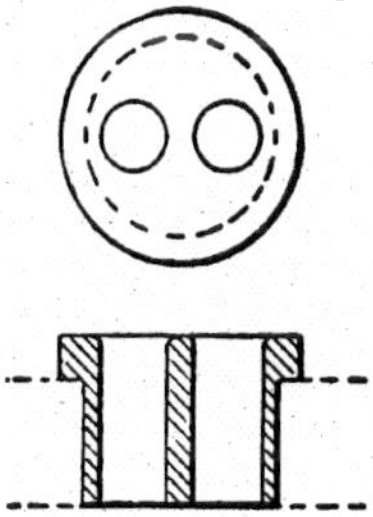

FIG. 54.

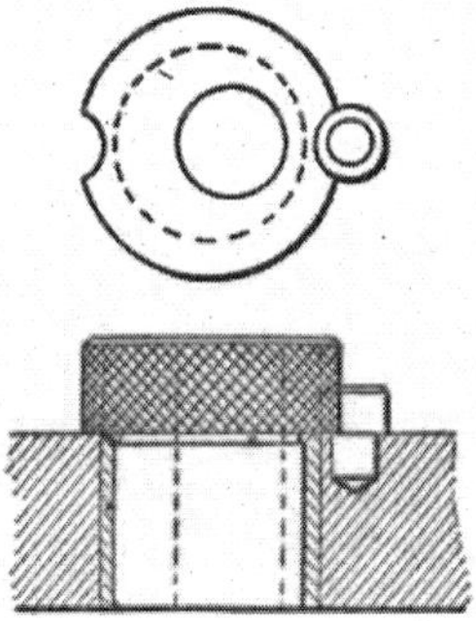

FIG. 55.

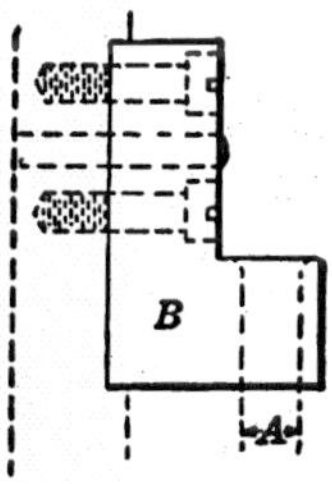

FIG. 56.

FIGS. 46 TO 56.—DIFFERENT TYPES OF JIG BUSHINGS.

BUSHING FOR EXTENSION DRILL

Another design of stationary bushing is frequently used when a bushing is too long to allow the use of a standard drill. In this case the drill is made into what is called an extension drill. This bushing, as shown in Fig. 48, will allow the extension part of drill to enter into it.

TWO PART BUSHINGS

A stationary bushing similar to the above except that it is made up of two parts, as in Fig. 49. The body *A* is generally made of soft steel or cast iron and the hardened bushing *B* for guiding the drill is inserted.

The design of bushing in Fig. 50 is similar to Fig. 48, except that it is used as a slip bushing and is guided by the stationary bushing *A*.

SLIP BUSHINGS

Standard slip bushings universally used in jig making for guiding a drill or reamer are made as in Fig. 51. Dimensions of these bushings are given in Table No. 17.

The illustration in Fig. 52 is simply shown to caution the designer or toolmaker to make the head *A* of the slip bushing always slightly larger than the outside diameter *B* of the stationary bushing. The reason for doing this is to prevent the stationary bushing from being forced out of the jig in case the operator should accidentally strike the head of the slip bushing.

To prevent the bushing being forced out of the jig some designers and toolmakers slightly taper the stationary bushing, as shown in Fig. 53, but this is not often necessary and is very seldom done.

BUSHINGS FOR TWO HOLES

It often happens that two or more holes are so close together that it is impossible to have independent bushings for each hole and these are then placed in one bushing, as shown in Fig. 54, or one hole is put in an extension bushing, as shown in Fig. 55. This bushing is then reversed for drilling the second hole.

BRACKET BUSHINGS

It is also found in many cases that a round bushing cannot be used at all and in such instance it is good practice to make the

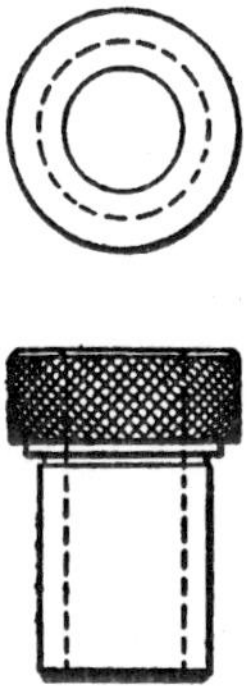

FIG. 57.

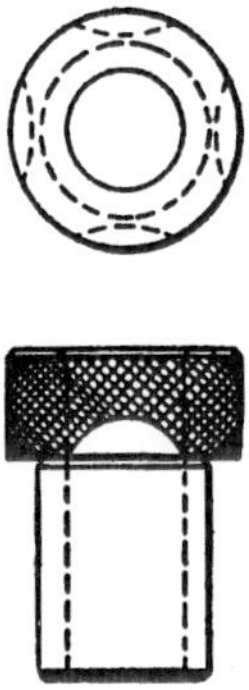

FIG. 58.

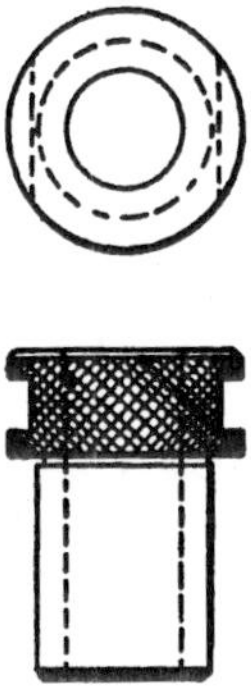

FIG. 59.

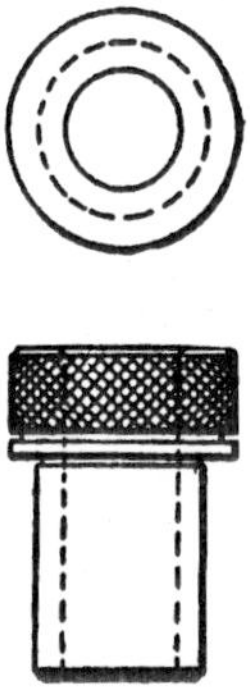

FIG. 60.

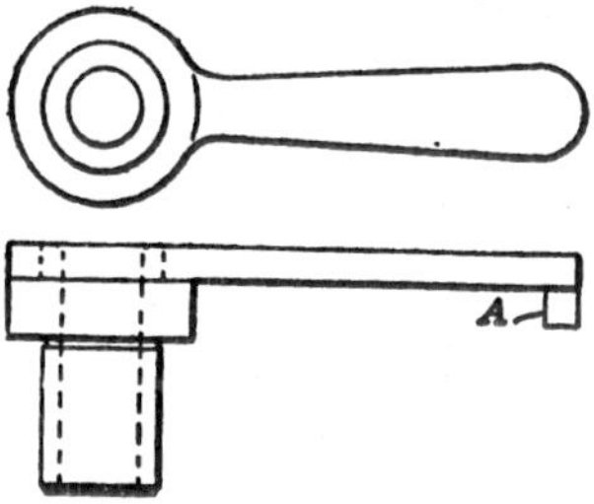

FIG. 61.

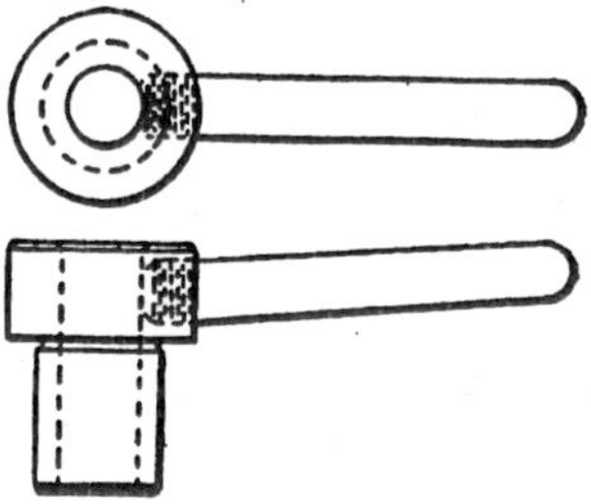

FIG. 62.

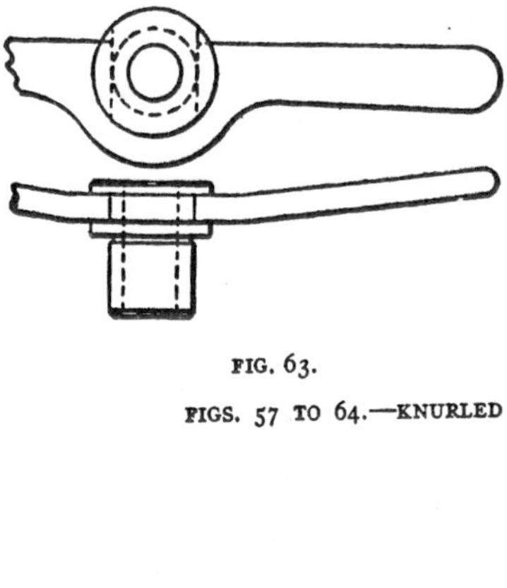

FIG. 63.

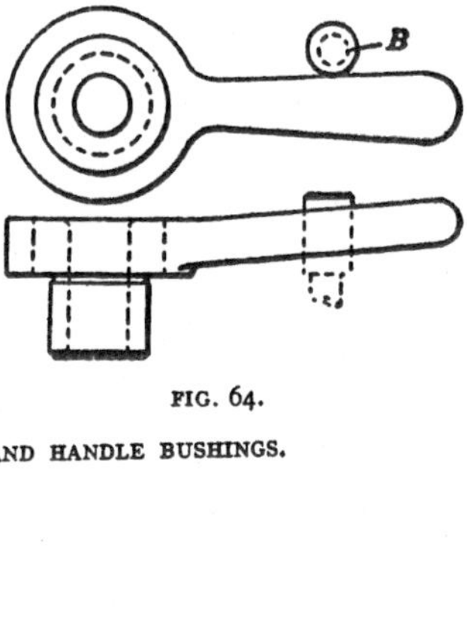

FIG. 64.

FIGS. 57 TO 64.—KNURLED AND HANDLE BUSHINGS.

bushing in the form of a bracket, as shown in Fig. 56. The hole *A* guides the drill or reamer and the body *B* is screwed on the wall of the jig and dowel pinned when in place.

REMOVING SLIP BUSHINGS

It is always necessary to provide means for removing slip bushings. The illustrations show a few schemes which work out well, Figs. 57 and 58 being generally used. This method permits the use of a screwdriver for removing the bushings and only applies to small sizes.

For the larger sizes handles can be attached to the head of the bushing as shown. In the best designs a pin *B* can be placed in the jig, as shown in Fig. 64, or a handle *A* can be bent, as shown in Fig. 61, to prevent the bushing from turning.

TO PREVENT BUSHINGS FROM TURNING

Another important feature which is too often overlooked is means for preventing bushings (especially small sizes) from turning or coming out. A few of these methods are shown below, Figs. 65 and 68 permit the bushing to turn but hold it in place, while Figs. 66, 67 and 69 hold it in place and also prevent the bushing from turning. Fig. 67 is the most commonly used.

SCREW BUSHINGS

Fig. 70 is a standard screw bushing which is used as a guide for a drill and also acts as a means to hold or help hold the work while being drilled, or it may also act as a guide by being cupped, as shown in Fig. 71. Table 18 gives dimensions of this style of bushing.

The illustration shown in Figs. 72 and 75 is the standard form of screw bushing containing a slip bushing. Fig. 72 shows the general practice when designing or making a bushing which is to simply act as a means for holding the work, and Fig. 75 when it is desired to hold and also guide the work by means of the cupped end. The dimensions are given in Table 18.

VERY ACCURATE BUSHINGS

If extreme accuracy is desired the bushings should be constructed as shown in Figs. 73 and 74. Fig. 73 is the style more commonly

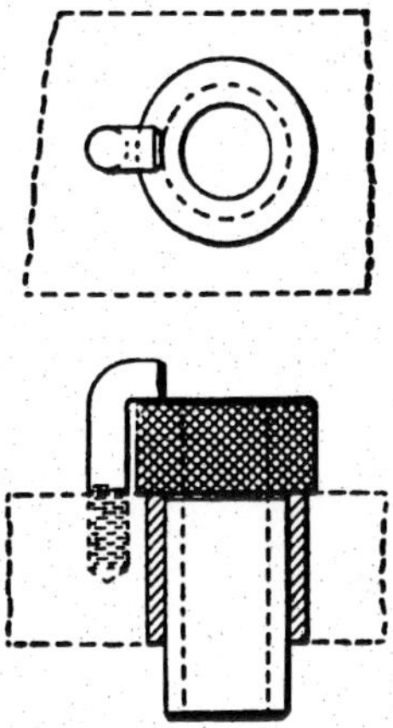

FIG. 65.

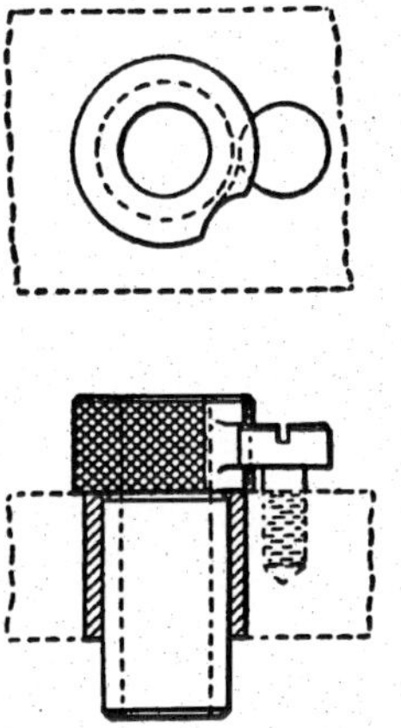

FIG. 66.

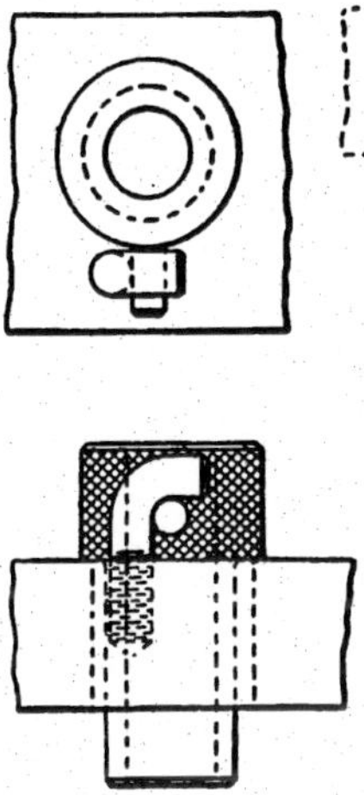

FIG. 67.

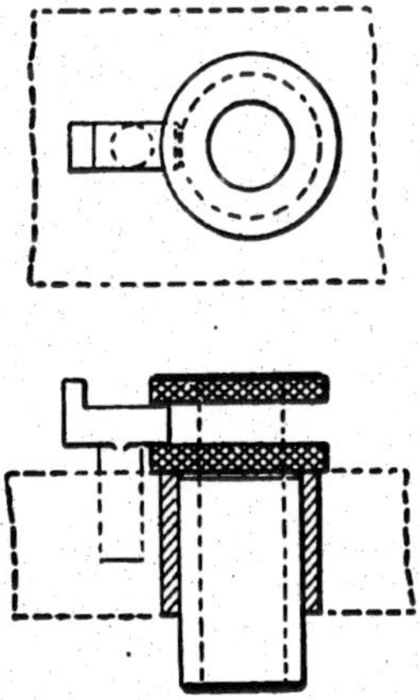

FIG. 68.

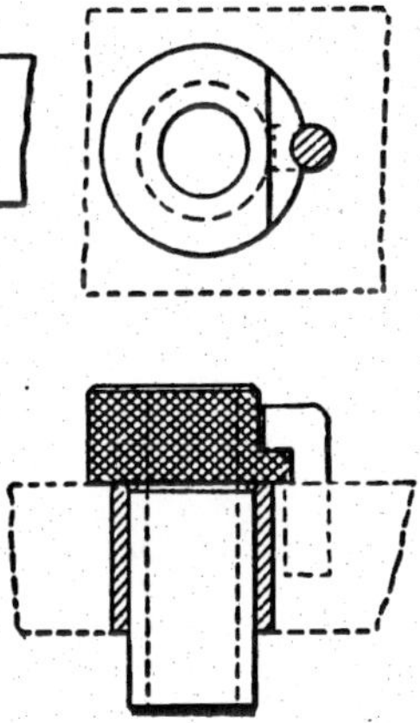

FIG. 69.

FIGS. 65 TO 69.—METHODS OF LOCKING BUSHINGS IN PLACE.

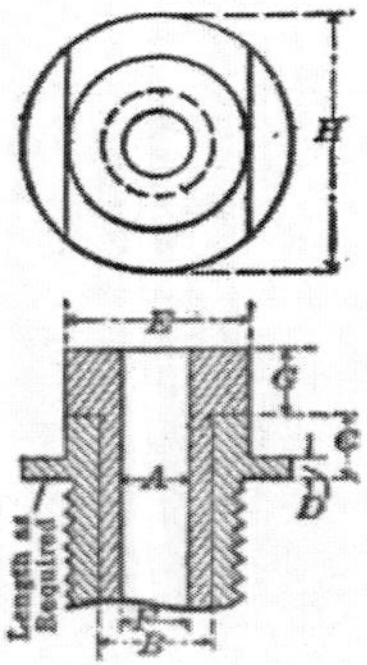

A	B	C	D	E	F	G	H	Thds.	A	B	C	D	E	F	G	H	Thds.
3/8"	1	3/8	1/8	1	5/8	3/8	1 1/2	12	1 1/4"	2 1/4	1/2	1/4	2	1 5/8	7/8	2 1/2	8
1/2	1 1/4	3/8	1/8	1	3/4	1/2	1 1/2	12	1 3/8	2 3/8	1/2	1/4	2 1/4	1 3/4	7/8	2 3/4	8
5/8	1 3/8	7/16	3/16	1 1/4	7/8	9/16	1 3/4	12	1 1/2	2 1/2	9/16	1/4	2 1/4	2	7/8	2 3/4	8
3/4	1 1/2	7/16	3/16	1 1/4	1	9/16	1 3/4	12	1 5/8	2 3/4	9/16	1/4	2 3/4	2 1/8	7/8	3 1/4	8
7/8	1 5/8	1/2	1/4	1 3/8	1 3/16	5/8	2 1/8	8	1 3/4	3	5/8	1/4	2 3/4	2 1/4	7/8	3 1/4	8
1	1 7/8	1/2	1/4	1 5/8	1 5/16	3/4	2 1/8	8	1 7/8	3 1/8	5/8	1/4	3	2 3/8	7/8	3 1/2	8
1 1/8	2	1/2	1/4	2	1 7/16	3/4	2 1/2	8	2	3 3/8	5/8	1/4	3	2 5/8	1	3 1/2	8

SCREW AND SLIP BUSHING.

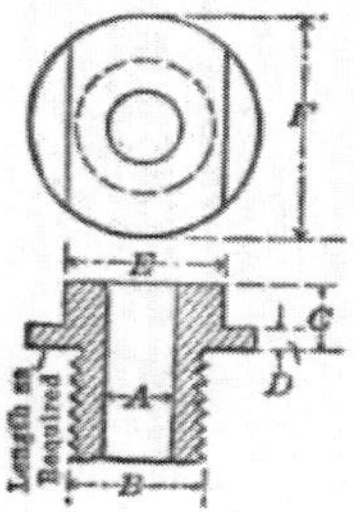

A	B	C	D	E	F	Thds.	A	B	C	D	E	F	Thds.
3/8	3/4	3/8	1/8	7/8	1 1/4	12	1 1/4	1 7/8	1/2	1/4	1 5/8	2 3/8	8
1/2	7/8	3/8	1/8	7/8	1 3/8	12	1 3/8	2	1/2	1/4	1 3/4	2 1/2	8
5/8	1	7/16	3/16	1	1 1/2	12	1 1/2	2 1/8	9/16	1/4	2	2 5/8	8
3/4	1 1/8	7/16	3/16	1	1 1/2	12	1 5/8	2 1/4	9/16	1/4	2	2 3/4	8
7/8	1 3/8	1/2	1/4	1 1/4	1 3/4	12	1 3/4	2 3/8	5/8	1/4	2 1/4	2 7/8	8
1	1 1/2	1/2	1/4	1 1/4	1 3/4	12	1 7/8	2 1/2	11/16	1/4	2 3/8	2 7/8	8
1 1/8	1 3/4	1/2	1/4	1 5/8	2 1/4	8	2	2 5/8	3/4	1/4	2 1/2	3 1/4	8

STANDARD SCREW BUSHING.

TABLE 18. TWO KINDS OF SCREW BUSHINGS.

used. In either case the thread should be an easy fit to allow the pilot of bushing *A* to have a sliding fit as this should act as the guide rather than the thread.

In this construction it is well to have a hardened and ground stationary bushing for them to work in, so that in case of wear

FIG. 70. FIG. 71. FIG. 72.

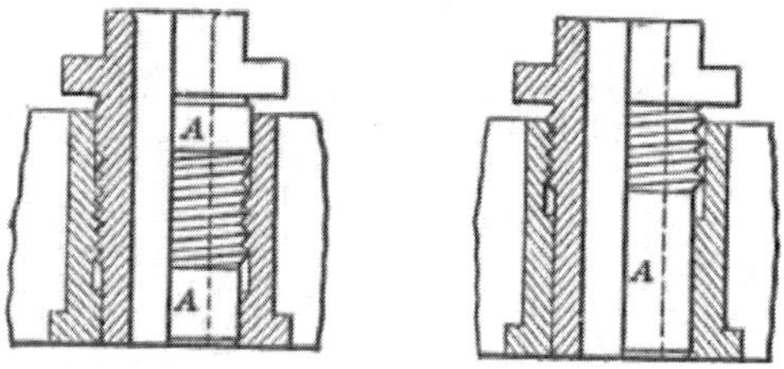

FIG. 73. FIG. 74. FIG. 75.

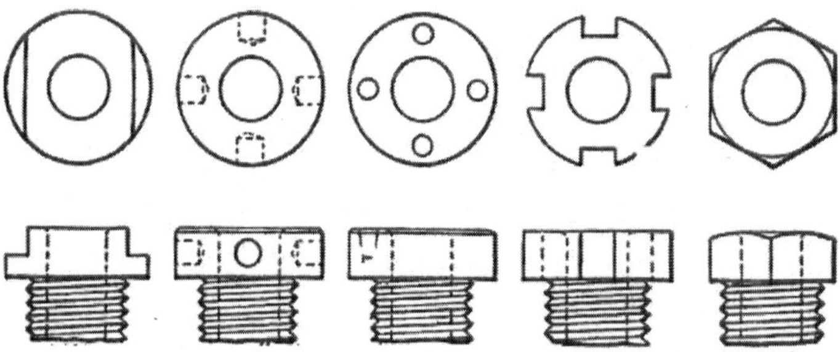

FIG. 76. FIG. 77. FIG. 78. FIG. 79. FIG. 80.

FIGS. 70 TO 80.—SCREW BUSHINGS OF DIFFERENT TYPES.

they can be removed. This is good practice for all styles of screw bushings.

For operating the screw bushings, it is customary to use a monkey wrench, although special wrenches can be made up as shown. These are used in conjunction with the screw bushings shown. Figs. 76 and 77 is the style head most commonly seen. Figs. 78, 79 and 80 show other forms of heads on screw bushings.

TREATMENT OF MACHINE AND COLD-DRAWN STEEL FOR BUSHINGS

Upon the completion of the bushings by the lathe hand they are sent to the case-hardening department to be carbonized.

After they are carefully packed in bone in cast-iron pots, which are sealed with fire clay, they are then placed in the furnace, which is kept between 1600 and 1700 degrees Fahrenheit, and left in from 4½ to 5 hours. The carbon will then have penetrated about 1/32 in.

The pots are then taken out of the furnace and allowed to cool off, after cooling they are unpacked and the bushings again placed in the furnace, heated to 1550 degrees Fahrenheit and quenched in a solution of salt and water.

TREATMENT OF TOOL STEEL FOR BUSHINGS

Tool-steel bushings are heated in a lead bath to 1550 degrees Fahrenheit, then quenched in water and cooled off in oil; after tempering they are drawn to 440 degrees Fahrenheit. This applies especially to Jessops and Saranac tool steels, but similar treatment can be used for other brands.

CHAPTER V

SOME OF THE DETAILS OF JIG MAKING

Drill jig lids may be divided into two classes, removable and swinging. There are a great many ways of designing them and many points in this construction are too often overlooked.

Figs. 81, 82, 83 and 84 are good examples of everyday practice in designing or making removable lids. Figs. 81 and 82 are very similar. Where extreme accuracy is to be maintained for a very great length of time, Fig. 81 answers nicely. Hardened plugs are driven in the jig body and lid. They accurately locate the lid as well as prevent wear.

On large jigs, means should be provided for removing the lid. In some instances an eye bolt *B* is provided as shown in the illustration.

THE DIRT GROOVE

In Fig. 82 will be found the general practice and attention here is called to the dirt groove which is so often overlooked, and it will be found of advantage to rib these lids whenever possible, either on top or bottom.

Figs. 83 and 84 are very much alike. In Fig. 83 the stud is made large enough to allow the lid to be removed without taking off the nut. The nut is simply loosened and the slip washer removed. In Fig. 84 the lid is located by means of the two studs similar to Fig. 83, but to remove the lid, the two eye bolts are released and swung out of the way. It is always advisable to harden the studs in both cases and provide hardened bushings in the lid for them to work into. As a word of caution do not make the studs any longer than is absolutely necessary, as the longer the studs the more difficult the operation of removing the lid.

Figs. 85 to 92 show a few examples of swinging lids. Fig. 87 is an improvement over Fig. 85, on account of the hardened-steel bushings. Its wearing qualities are doubled and its accuracy maintained.

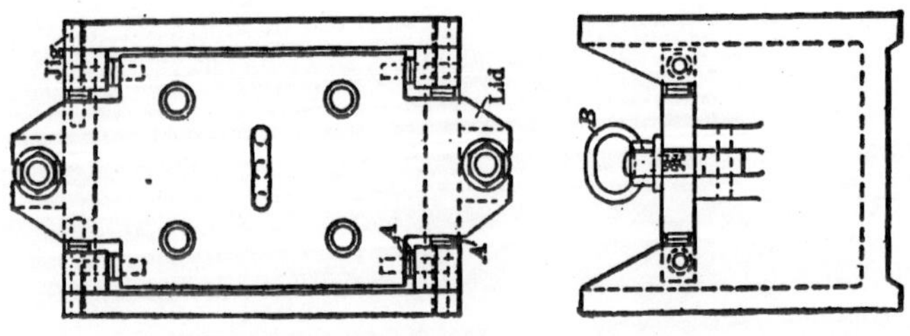

FIG. 81.

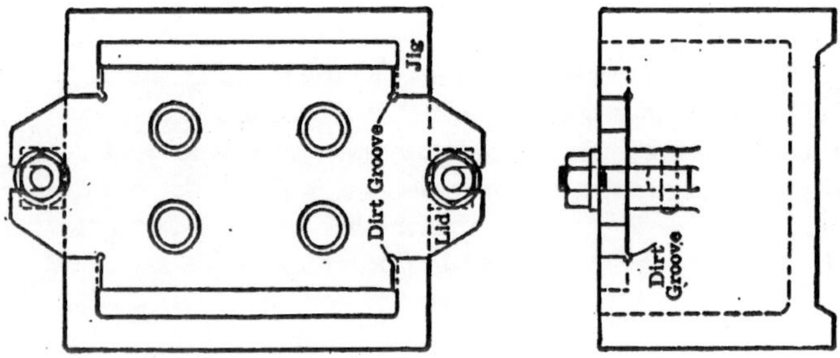

FIG. 82.

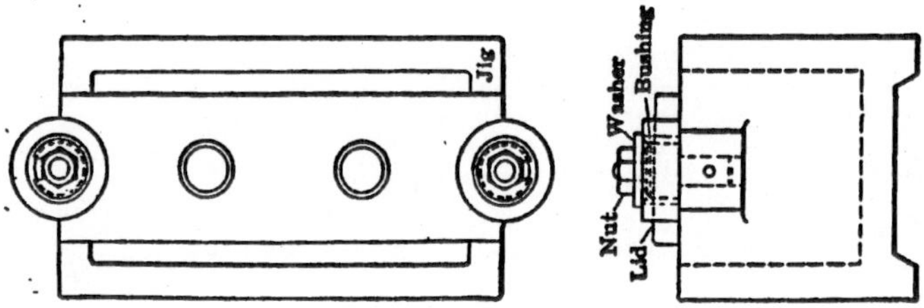

FIG. 83.

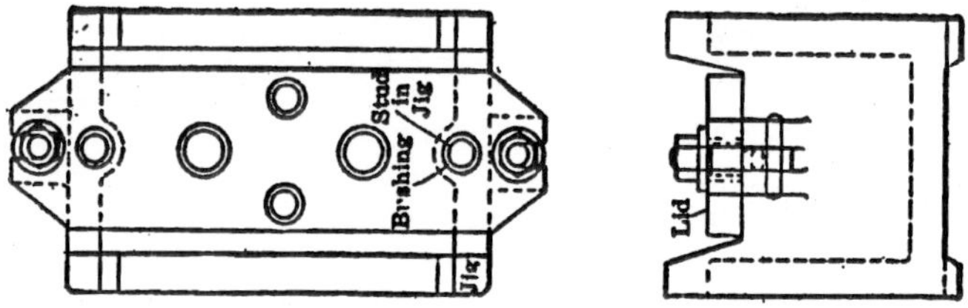

FIG. 84.

FIGS. 81 TO 84.—DETAILS OF DRILL JIG LIDS.

Fig. 88 is an improvement over Fig. 86 and is probably the most commonly used.

Fig. 89 illustrates a method of securing the opposite end of a swinging lid and takes the place of the eye bolt as shown in Fig. 90. Hardened-steel bushings are placed in the jig body and in the lid.

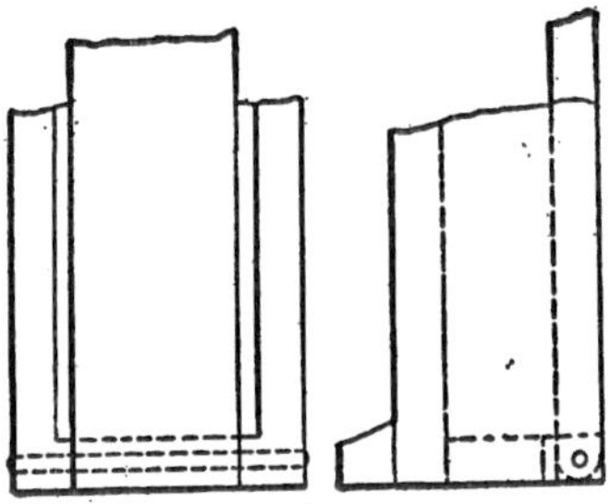

FIG. 85.

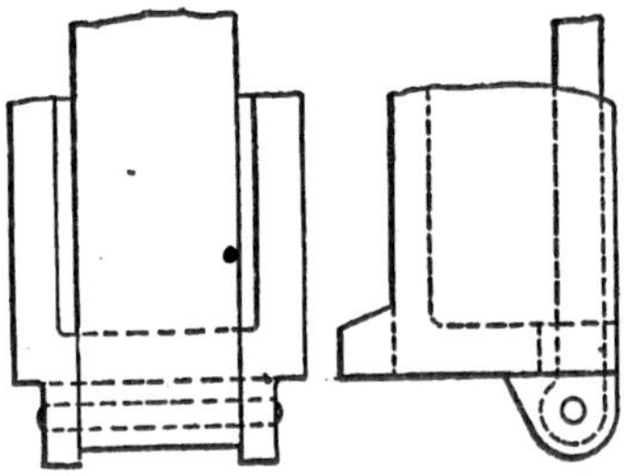

FIG. 86.

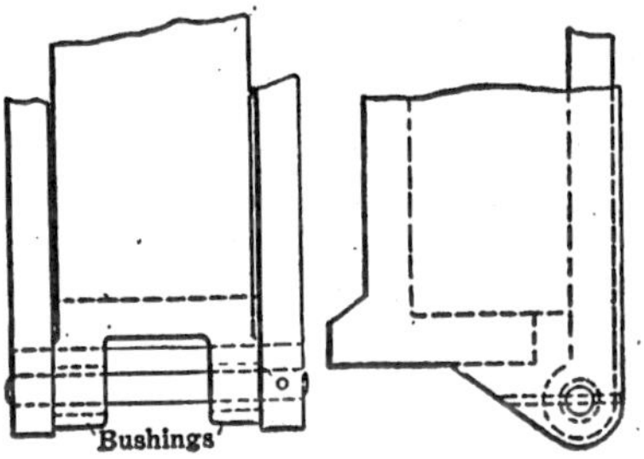

FIG. 87.

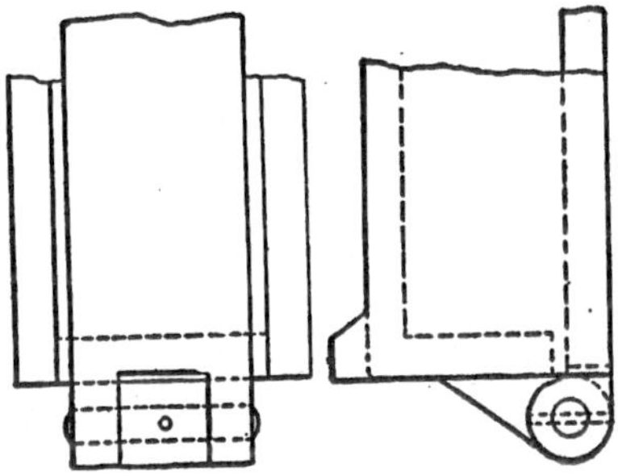

FIG. 88.

FIGS. 85 TO 88.—TYPES OF HINGED LIDS.

The pin being a sliding fit holds the lid properly. Attention may be called to the designing of the lug in Fig. 90. The lug acts as a support for the bolt and seat for the lid. In this construction there is no danger of the lid being broken or sprung out of shape in case the operator puts an excessive amount of strain on the bolt.

Fig. 91 is made similar to Fig. 81, it having the inserted studs *A* for keeping the lid in alinement and also increasing its wearing qualities.

Fig. 92 is an improvement over Fig. 87. The bushings are so designed to take up the wear on the pin as well as the sides of the

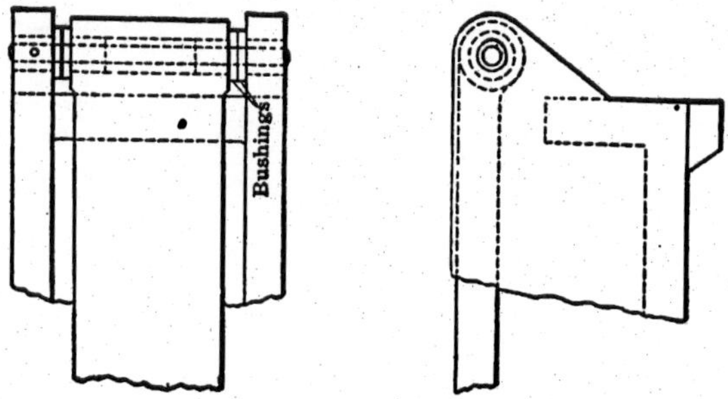

FIG. 92.

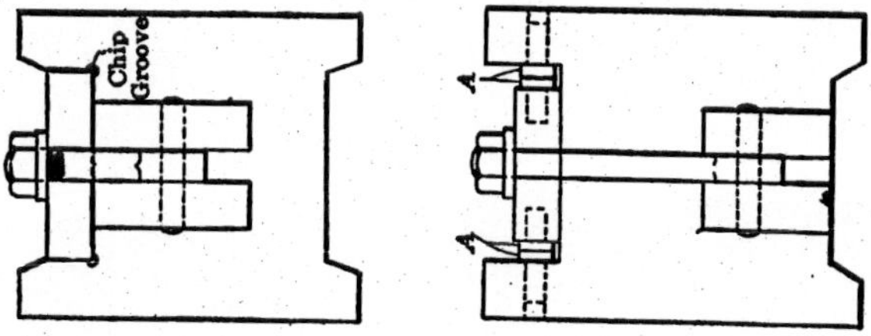

FIG. 91.

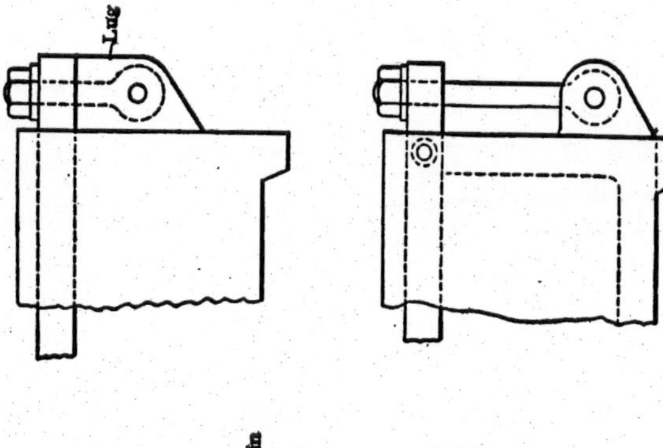

FIG. 90.

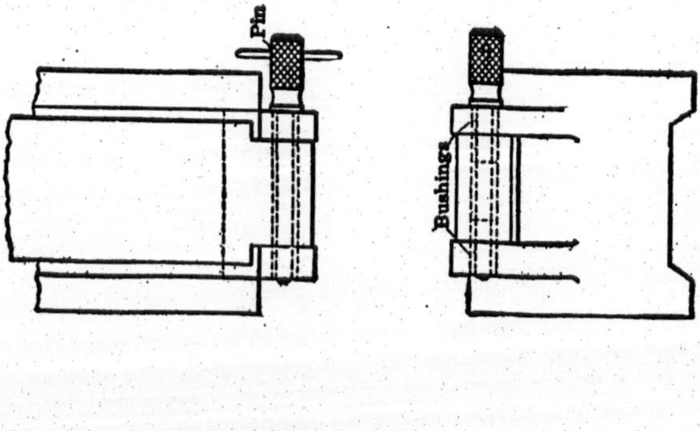

FIG. 89

FIGS. 89 TO 92.—METHODS OF LOCKING JIG LIDS.

lid. This design is expensive and only used where continued accuracy is required.

JACK SCREWS AND PINS

Figs. 93, 94, 95 and 96 are examples of jack screws which are used for supporting the part to be drilled. They are also used for taking up the thrust of the drill. Figs. 93 and 96 are the most commonly used style of screws. The jack screws are, however, in many instances, superseded by the jack pins, for the reason that the jack screws must be adjusted very carefully to prevent springing or distorting the work, and chips, oil or dirt working its way into the threads prevents it being operated quickly, etc.

While the jack screw has its disadvantages, so has the jack pin, but these disadvantages can always be attributed to their design. A number of illustrations showing different methods of making them will be found in Figs. 97 to 104 inclusive. It is unnecessary to explain the operation of all of them as they speak for themselves, but a word of caution to the designer or maker is essential.

Fig. 97 is not protected from chips, powdered metal or dirt entering the sides of the pin. This will, therefore, be apt to soon get out of order and the pin stick in the hole provided for it.

Fig. 98 is protected from the chips, dirt, etc., by means of a cap. It is operated by a spring and held in position by the screw which strikes the tapered part of pin.

Fig. 99 is operated by means of a cone-pointed hardened screw. The illustration shows the screw having a square head but in most cases it has a knurled or winged head. This design is apparently correct but it has some bad features.

The power of a screw is too often misjudged by the designer as well as by the machine operator who is to use the device. Here we have the powerful action of the screw as well as the wedge action of the conical end of the screw against the inclined surface of the plunger pin, therefore very little pressure on the screw is sufficient to spring a good-sized casting several thousandths of an inch.

Fig. 102 is similar to Fig. 99, but in this case it is operated by means of a wedge-shaped pin which in turn is operated by a spring. The illustration shows the plunger pin at its extreme height. To loosen the plunger pin, the wedge-shaped pin is pulled out as far as it can be and the knob turned half way around. In so doing the

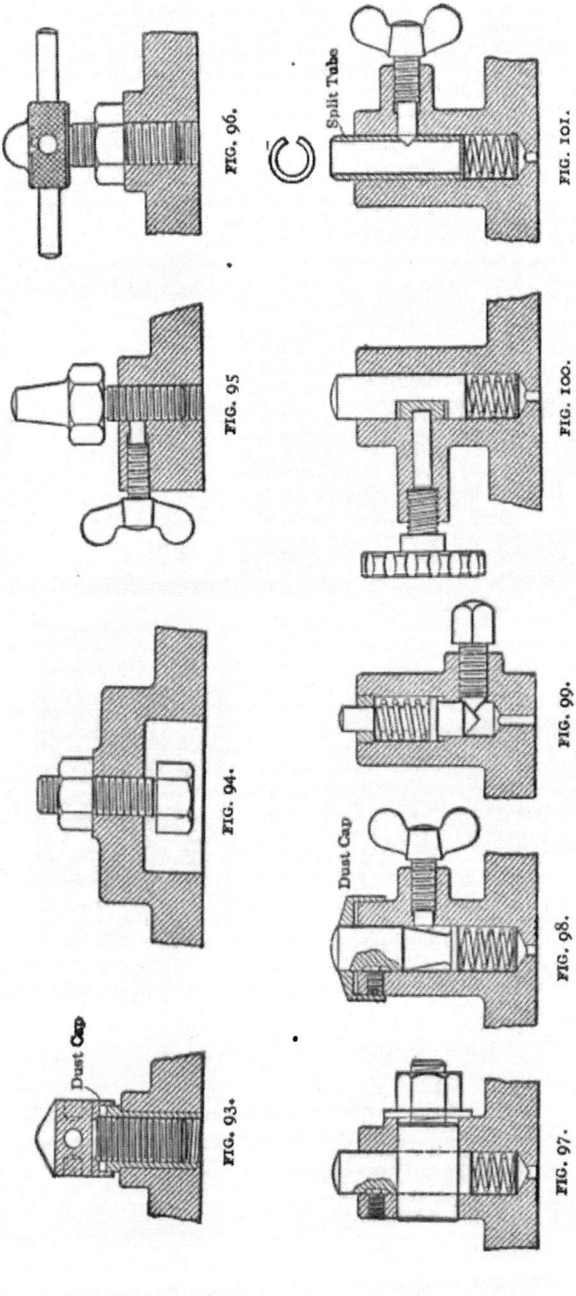

FIGS. 93 TO 101.—A VARIETY OF JACK SCREWS AND PINS.

bottom flange of the knob will rest on the head of the small pin *A* which will hold it out and allow the plunger pin to drop.

Figs. 103 and 104 show two ways of operating and holding two plunger pins.

Jack screws are generally made of cold-drawn steel and case hardened. Jack pins are usually made of tool steel and hardened, although if made of cold-drawn steel and properly case hardened their wearing quality is very high. Always provide dust caps where possible.

DRILLING ANGLE AND BLOCKS

It is, of course, essential to have the drill jig as light as possible, making the designing of it (in case a hole is on an angle or out of square with other holes) very difficult in many cases. For this reason we have what is called an angle block, such as is shown in Fig. 105.

This block is generally made of cast iron and provided with legs in the bottom, similar to the drill jig. The illustration shows a jig resting in the block.

An adjustable drilling angle is shown in Fig. 106. One or two of these drilling angles in a large factory are sufficient to take the place of a great number of angle blocks.

The plates *A* and *B* may be made of cast iron, 7/8 in. thick, $C = 3/8 \times 1\frac{1}{4}$ cold drawn steel, $D = 1/2$ in. bolts. The greatest angle of *G* is about 75 degrees and the smallest 30 degrees. These drilling angles should be made up in two sizes for the large size $F = 8$ and $E = 12$ ins. and for the small size $F = 6$ and $E = 10$ ins.

STATIONARY OR ADJUSTABLE STOPS

Ninety per cent. of the drill jigs designed and made have some form of stops. They may be stationary or adjustable. These stops, or, as they are sometimes called, locating points, are used for properly locating the piece in the jig so that the holes will when drilled be in the right position. Figs. 107, 110 and 111 are permanent or stationary stops and Figs. 108 and 109, adjustable stops.

It is always good practice to make these stops stationary whenever possible as the adjustable ones very often work loose and result in spoiled work. They also furnish a good excuse for the operator who was careless in placing the part in the jig. For if the holes are

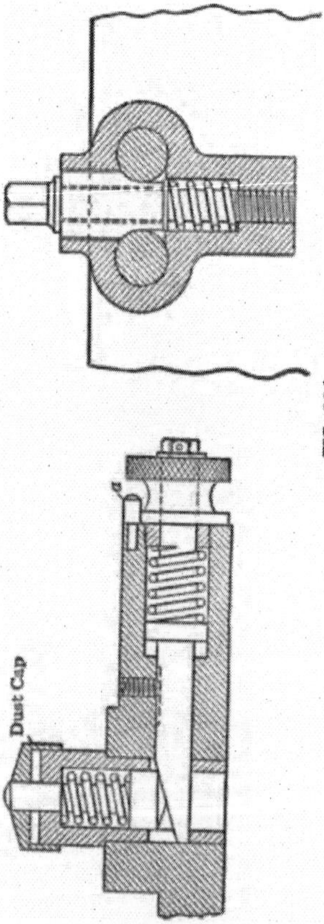

FIG. 102.

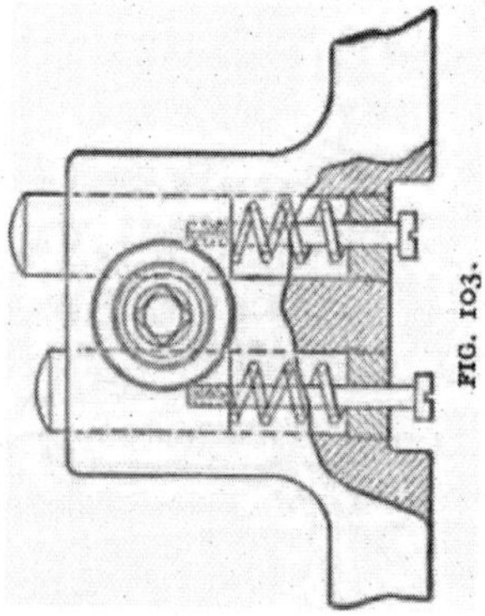

FIG. 103.

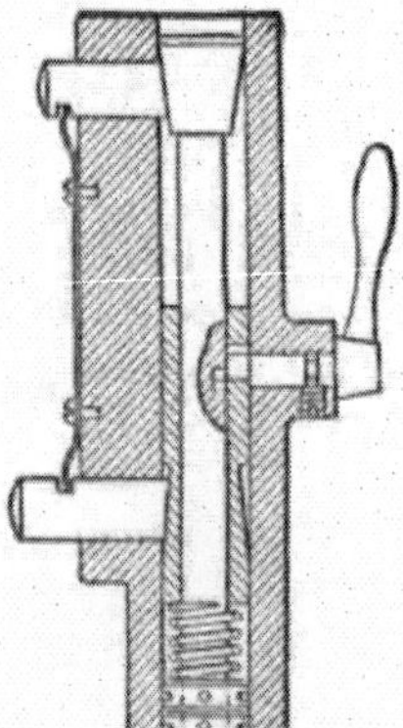

FIG. 104.

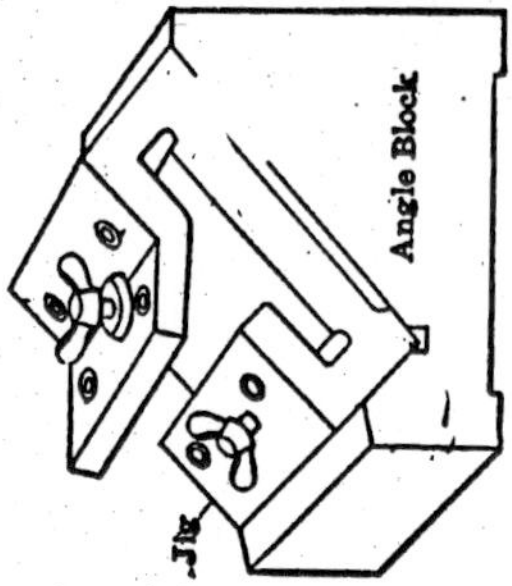

FIG. 105.

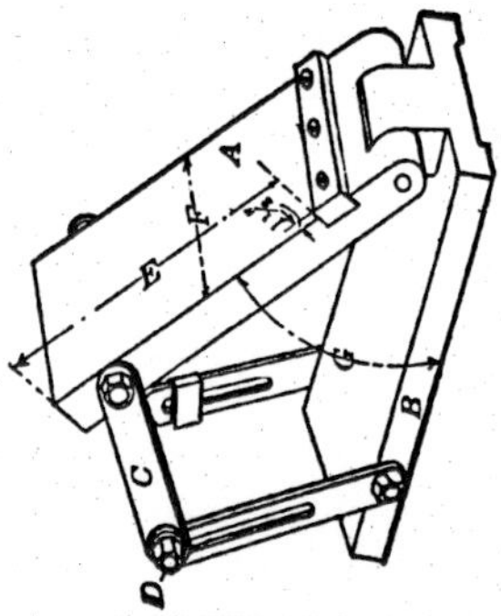

FIG. 106.

FIGS. 102 TO 106.—MORE JACK PINS AND TWO ANGLE PLATES.

drilled so that the part is spoiled, he can easily reset the adjustable stops and blame an innocent party.

Fig. 111 is a form of stop screwed and dowel pinned to the jig. This style is made up in many different ways.

They are usually made of tool steel, and hardened. In some

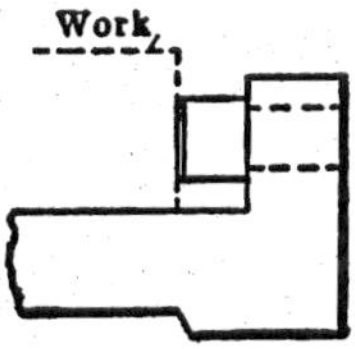

FIG. 107.

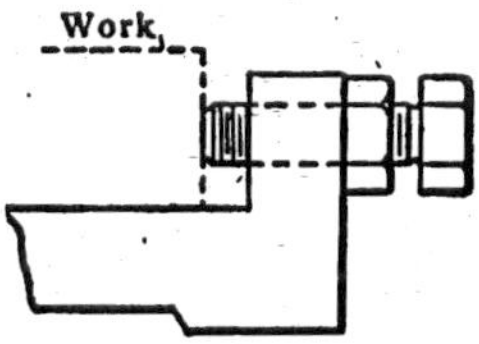

FIG. 108.

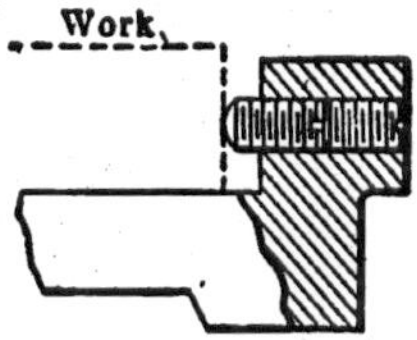

FIG. 109.

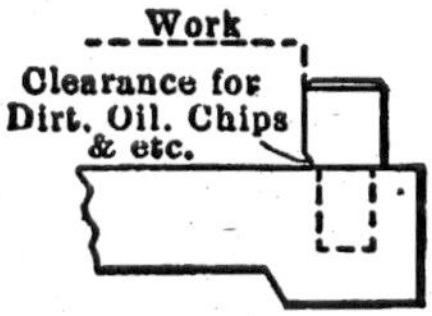

FIG. 110.

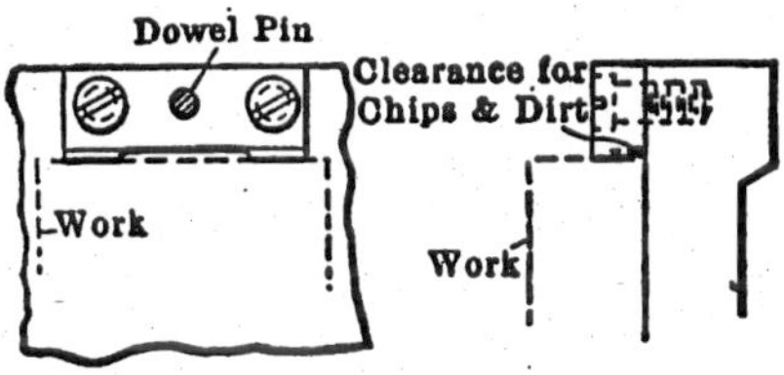

FIG. 111.

FIGS. 107 TO 111.—STATIONARY AND ADJUSTABLE STOPS.

cases they are made of cold-drawn steel and carbonized or case-hardened, and at times they are made part of the jig.

In designing all stops similar be sure to allow clearance for dirt, chips, etc.

INDEXING PINS AND METHODS

For the indexing of revolving or rotary drill jigs, the indexing method is governed by the design of the jig as well as the accuracy to be maintained in the spacing or location of the holes to be drilled. In the following illustrations will be found a few examples of indexing methods and the designer can select the one best suited for the work in hand.

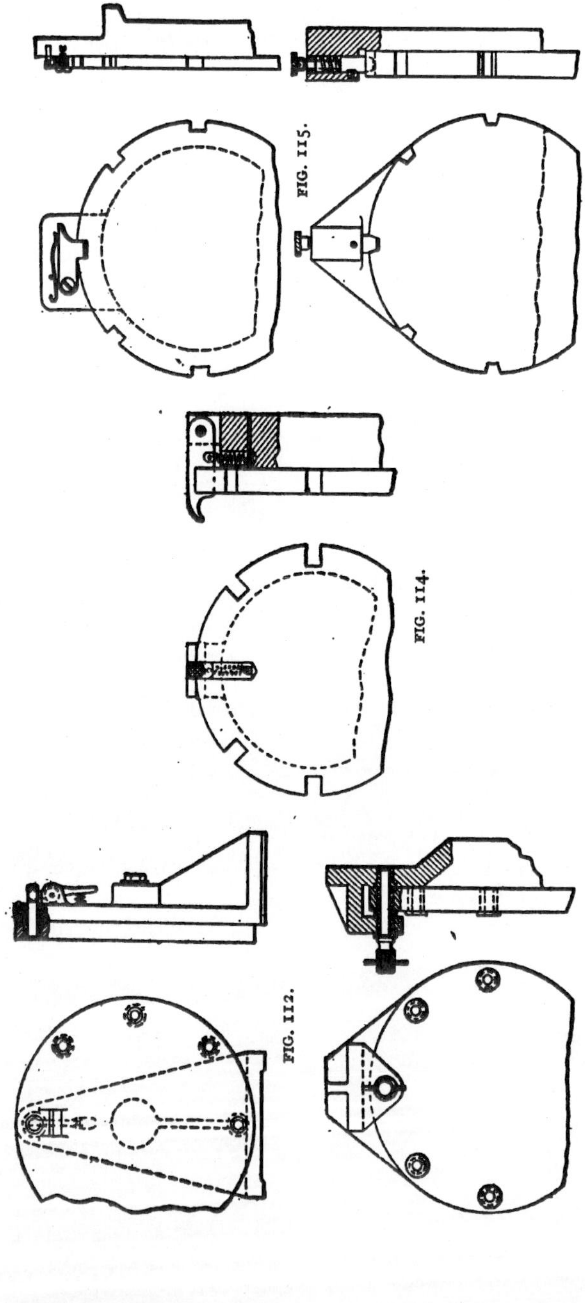

FIG. 112. FIG. 113. FIG. 114. FIG. 115. FIG. 116.

FIGS. 112 TO 116.—SEVERAL METHODS OF USING INDEXING PINS.

Figs. 112 and 113 will give the best results and may be used where extreme accuracy is required. The index pin in both cases makes this positive.

Fig. 112 has the advantage in not having to remove the index pin and, therefore, it can be operated more quickly. Many tool designers make the end of the index pin tapered and place corresponding bushings in the revolving member, but this has many

FIG. 117.

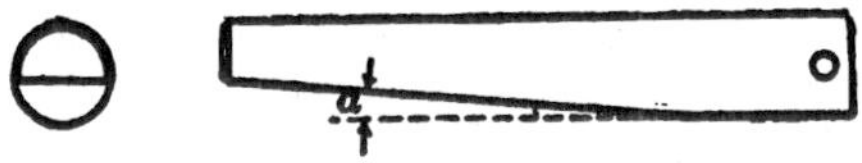

FIG. 118.

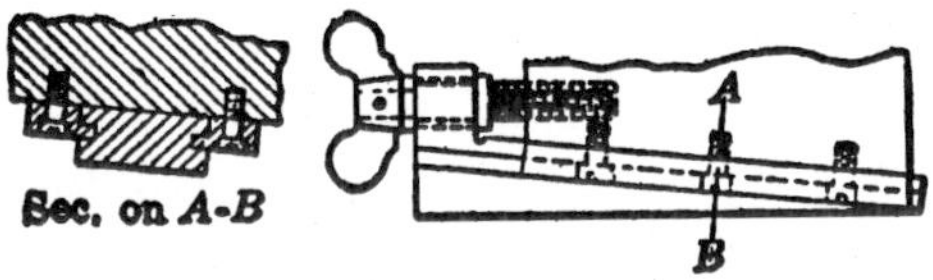

FIG. 119.

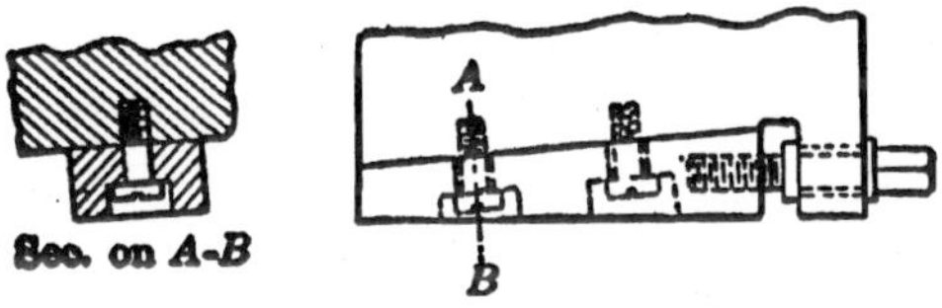

FIG. 120.

FIGS. 117 TO 120.—LOCKING WEDGES.

disadvantages. In the first place it would be absolutely necessary to design the jig so that there would be no danger of chips of iron or dust or dirt entering the tapered bushings, which would prevent the pin properly seating itself

Then, too, the compound generally used in the drill-press department soon works its way into the bearing which supports the revolving member and when the jig is not in use becomes hard or thick and thus prevents the member from freely turning, which causes considerable trouble in getting the pin to properly seat itself.

Figs. 114 and 115 are designs frequently used when there is not a great amount of accuracy required. They are equally good and in both cases the indexing slots and fingers are operated in a similar manner.

Fig. 116 probably originated from Figs. 112, 113 and 115. The index pin, however, is more durable than the index fingers.

WEDGES

The wedge is one of the simplest means of holding work in the drill jig known to the tool designer or maker, and therefore needs very little explanation.

They, like the cam, have a limited amount of movement, and should only be used where the variation in the size of the piece is within this amount.

Wedges are made up of square and round steel as shown in Figs. 117 and 118. The amount of taper *A* given them varies 4 or 5 degrees will give the best results.

Figs. 119 and 120 show two styles of wedges which are operated by means of a screw, and held in position with gibs or screws.

Formula for Cam Locks

Fig. 121 shows two types of cam locks. These diagrams, with the formula below, will show how to calculate and lay out locking for jigs and similar purposes.

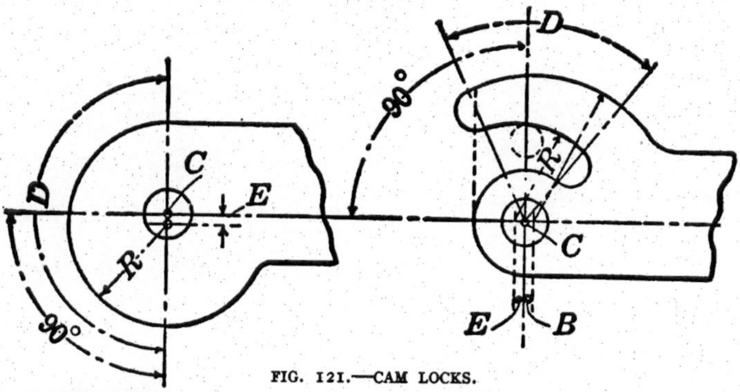

FIG. 121.—CAM LOCKS.

R = Radius,
C = Fulcrum of lever,
E = Amount of eccentricity = $0.001 \times 2 \times 90$,
$B = E$,
D = Number of degrees of total eccentricity. $0.001 \text{ in.} \times R \times D = 2 \times E$.

0.001 in. rise for every 1 degree when radius is 1 in.

In making these cams maintain a limit of 0.005 to 0.010. To lay them out, convert *E* into the nearest fraction and scribe radius *R*. The result will be close enough for jig work.

Example No. 1

What is the amount *E* and total amount *D* of eccentricity in a lever having a 1-in. radius?

0.001 in.
1 in. radius.
90 number of degrees.
0.090 in. = E
2
0.180 in. = D
0.001 in.
1 in. radius
0.001 in. rise of 1 degree
180 total number degrees
0.180
00.180 in. = D

Example No. 2

What is the amount of *E* and total amount *D* of eccentricity in a lever having a 1½-in. radius?

0.001
1.5 in.
0.0015 in.
90
0.1350 in. = E
2
0.2700 in. = D

0.001
1.5 in.
0.0015 in.
180
0.1200
0.15
0.2700 in. = D

CHAPTER VI

MISCELLANEOUS JIG DETAILS

Fig. 122 is a good example of applying what is sometimes called in tool designing a fool pin. The illustration shows a drill jig *A* with the part to be drilled *B* properly placed in the jig, and the fool-proof pin *C* which insures the piece being always placed in the jig in the right position.

It can be readily seen that should this pin be removed, the hole

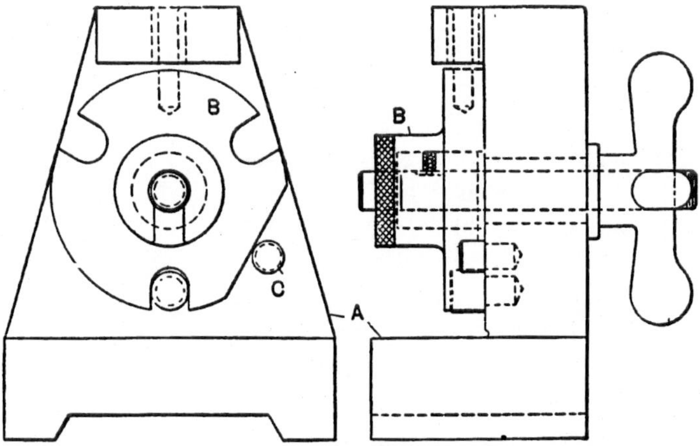

FIG. 122.—A PIN TO PREVENT MISTAKES.

could be drilled in three different points, leaving it to the judgment of the operator to place the piece in correctly.

Fig. 123 is a good example of applying a fool-proof clamp. By its use the same results are obtained as with the pin. The illustration in like manner shows the part *A* in a drill jig *B*. The clamp referred to is *C*, which allows the work to go in the jig but one way as well as holds it in place.

In the designing of jigs or fixtures, it frequently happens that methods for allowing the part to enter in the jig but one way are overlooked or left to the judgment of the operator. This is a serious mistake on the part of the designer, as a jig or fixture should be in all cases absolutely fool proof.

METHODS OF FASTENING BLOCKS, STUDS, ETC.

There are always more or less small parts such as shown in Figs. 124 to 129 inclusive, used in the building of tools, which are bolted

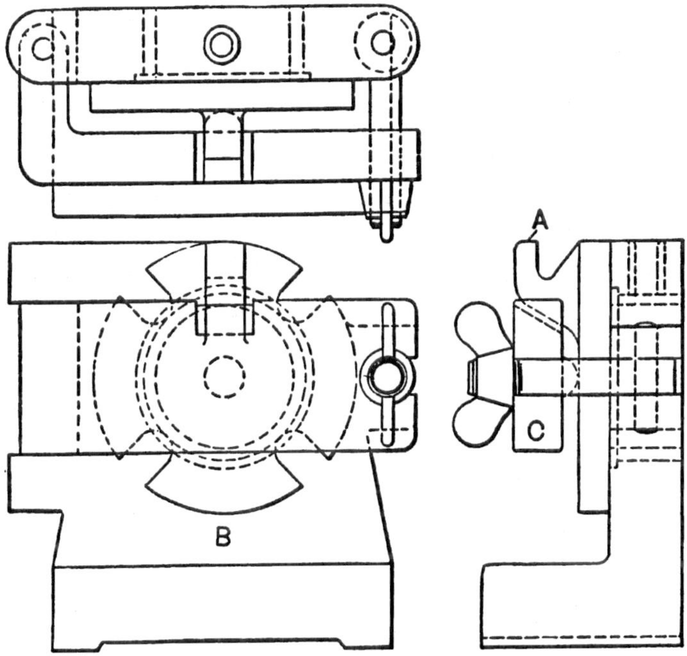

FIG. 123.—A CLAMP TO PREVENT MISTAKES.

or screwed to the jig in order to obtain efficient results for an indefinite period.

These parts may be subject to wear, in which case they are easily replaced. They are frequently used to hold drill bushings, binding screws, etc. In case of a change in the design of the part to be drilled, these extra pieces are used to hold additional drill bushings.

The illustration shows the regular practice of bolting and attention is called to the important point of properly dowel pinning them in place.

Fig. 124 is held by two cap screws and a key, while Fig. 125 is held by a dowel pin instead of a key.

Fig. 126 is similar to Fig. 125, the only difference being the

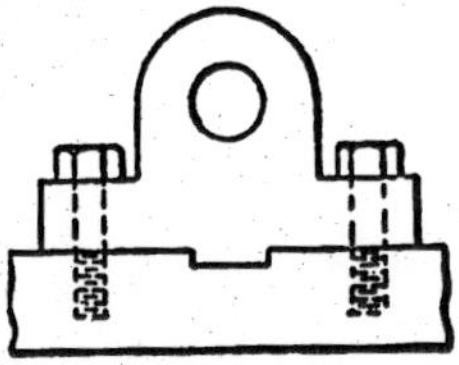

FIG. 124.

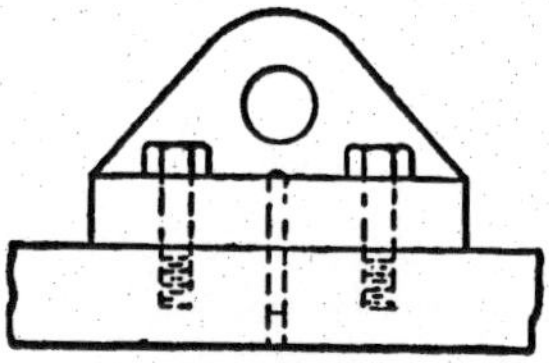

FIG. 125.

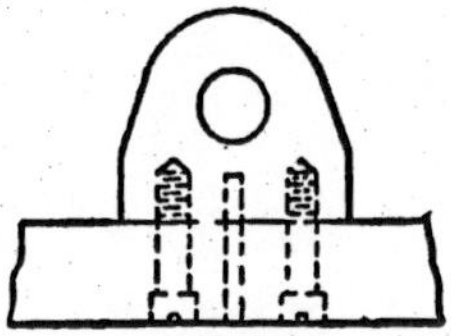

FIG. 126.

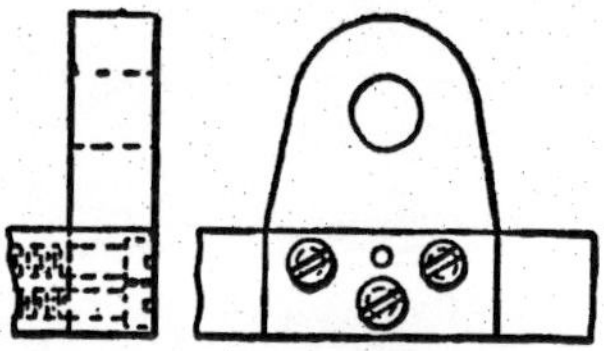

FIG. 127.

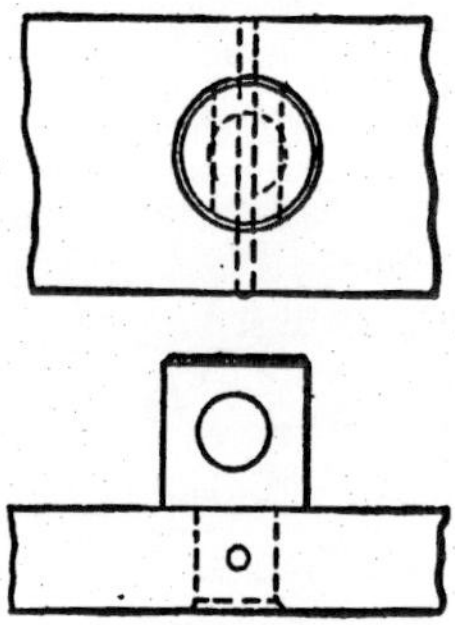

FIG. 128.

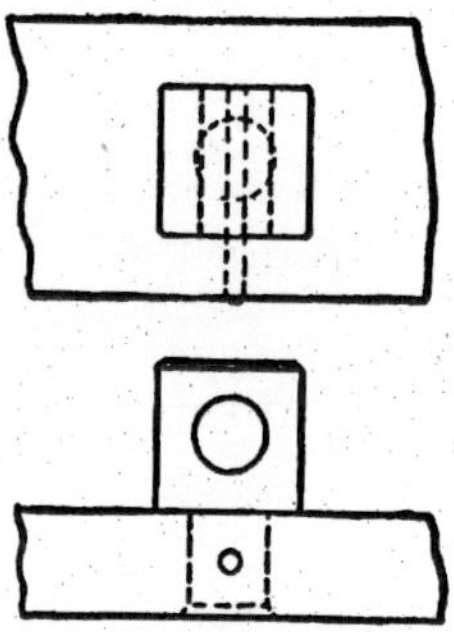

FIG. 129.

FIGS. 124 TO 129.—FASTENING BLOCKS AND STUDS.

shape of the part and the fillister-head screw which takes the place of the bolts, or cap screws.

Fig. 127 shows a practical way of securing a part to the side of the jig or fixture.

Figs. 128 and 129 illustrate the use of a round or square stud. The shanks of these studs are turned down and driven in the hole provided for them, after which they are riveted and pinned in place. In some instances their shanks are threaded and they are screwed in the jig or fixture.

JIG PROFILES

One of the most difficult things in tool designing is providing means for properly locating the piece in the jig.

This is experienced more where there is a number of irregular bosses and it is more essential to have the holes come in the middle

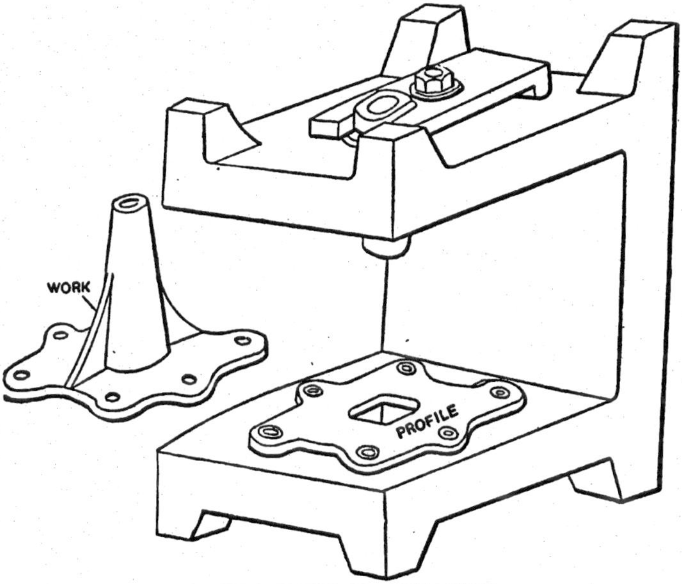

FIG. 130.—USING PROFILE TO LOCATE WORK.

of these than to have them accurately spaced in relation to one or more holes in the same part. One of the most useful means in a case of this kind is what is called a profile on the base of the jig, such as is shown in Fig. 130. This profile consists of a piece of sheet steel about 1/8 in. thick and is cut to correspond to the flange or base of the piece, after which it is screwed as well as dowel pinned on the base of the jig.

In placing the piece in the jig, it is set upon the profile which will readily take care of locating it, after which it is clamped down and drilled.

Fig. 131 is a cast-iron plate which contains a number of drill bushings. It is located and held in position by the studs *A A*.

This makes a quick method of removing a number of drill and reamer bushings.

The illustration shows the plate carrying the drill bushing.

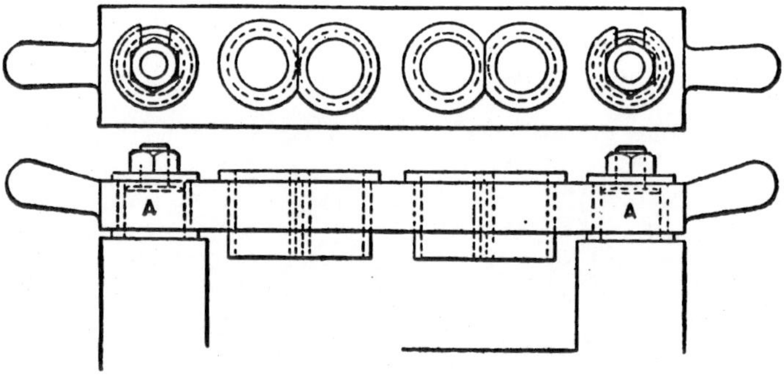

FIG. 131.—A QUICK WAY TO REMOVE BUSHINGS.

A second plate, practically the same (except that reamer bushings are used instead of the drill bushings), is not shown.

It is also used when necessary to drill and ream holes which are very close together as the illustration shows.

ADJUSTABLE SCREW AND SLIP BUSHINGS

In many cases it is more essential to have the hole drilled in the center of the boss, than it is to drill it in any particular relation to other holes in the same part. For this reason we have what may be called an adjustable screw or slip bushing, such as is shown in Figs. 132 and 133.

The sliding blocks or bushings *A* carry the drill bushing. These blocks or bushings are made to slide longitudinally in a hole or groove provided for them in the jig.

It is absolutely necessary that they slide freely as well as have no perceptible shake and where possible a chip guard should be used to prevent chips, dirt, etc., from interfering with their movement.

These adjustable bushings are made up in many different ways; the two examples shown are most commonly used.

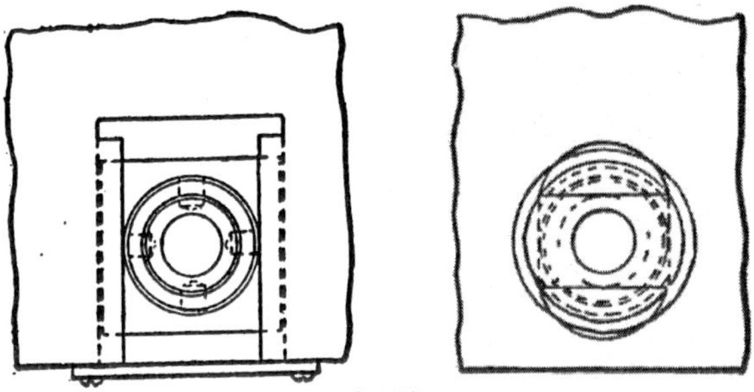

FIG. 132.

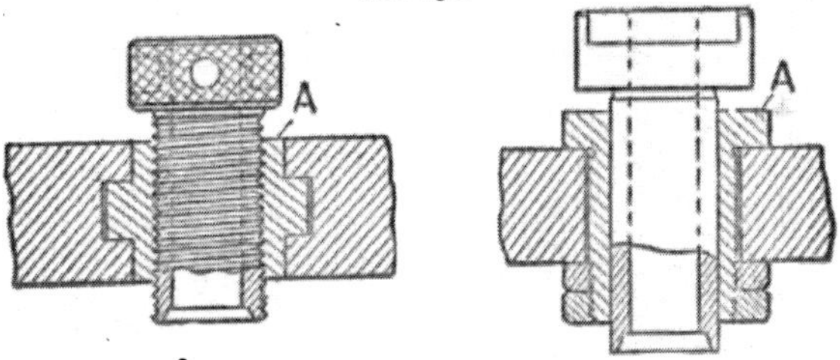

FIG. 133.

FIGS. 132 TO 133.—ADJUSTABLE SCREW AND SLIP BUSHINGS.

GAGES FOR USE IN DRILLING

Fig. 134 shows a way of gaging the depth of one or more counterbored holes or spot-faced bosses and shows the part *A* in the jig *B*. The holes in the piece have been counterbored and the plugs *C C* are shown in position, ready for gaging the depth of them. The

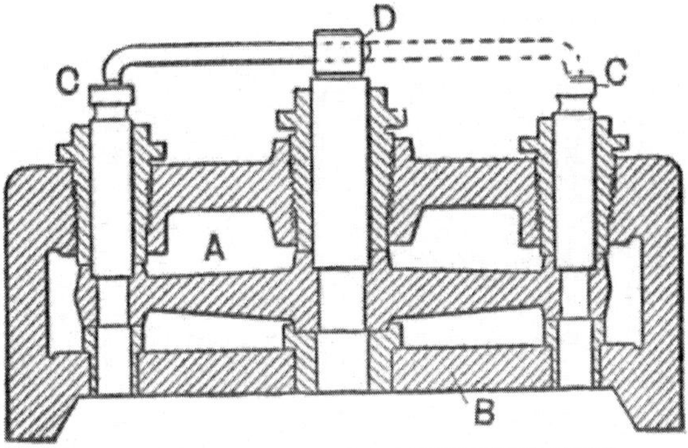

FIG. 134.—GAGING DEPTH OF COUNTERBORED HOLES.

hight gage *D* rests in the center counterbored hole and is swung from one plug to the other, readily detecting any variation.

In some instances these hight gages are made with a flush pin gage on the end of the rod, in place of the solid end, and when very accurate results are necessary an indicator is attached.

Fig. 135 is a small depth gage, such as is used in the drill-press department for determining the depth of a drilled or counterbored

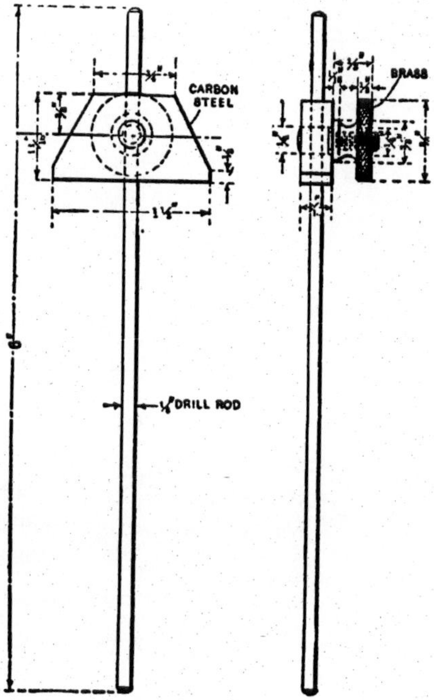

FIG. 135.—A GAGE FOR USE IN DRILLING.

hole. It is also used when spot-facing bosses, this being frequently done while the part is still in the jig.

A variety of thumb screws and handles to be used in jig work of different kinds will be seen in Fig. 136.

POINTS TO CONSIDER IN DESIGNING JIGS

What kind or class of jigs are you going to design? Will any of the standard jig designs shown help you?

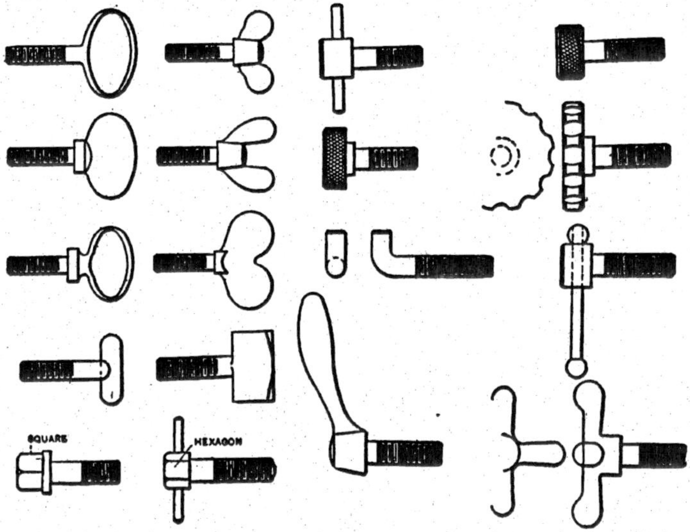

FIG. 136.—THUMB SCREWS AND HANDLES FOR JIGS.

Has the part undergone any previous operation? If so, can you utilize any of these points to start from?

Is it absolutely necessary to work from any previous finished or bored operation?

If locating against rough or unfinished surface, is it advisable to have locating points adjustable?

Can any of the clamping schemes shown be used to hold the part securely while it is being drilled and reamed?

Can any of the standard stock castings or patterns shown be used in making the jig, or must you make a special pattern?

Can the work be held down by any of the methods illustrated? If so, will the results after drilling and reaming be accurate enough?

What takes the thrust of the drill? Can you use any jack pins or screws to support the work while it is being drilled?

In the building up of the jig, are all parts properly screwed and dowel pinned in place? Is there any danger of any of these parts working loose?

Is the jig fool proof so that when the piece is inserted it will always go in the correct way?

Can the piece be quickly removed?

Can the chips be quickly brushed away?

Is it necessary for the piece to rest on more than three locating points in jig?

Can the locating points in jig be readily cleaned and kept free from dirt, oil, chips, etc.?

Is the jig strong enough?

Have the stationary bushings plenty of bearing?

Can any of the standard screws and slip bushings shown be used?

Is it necessary to have slip bushings?

Are the bushings close to the work?

Are they at least twice as long as the diameter of the drill?

Are the heads of the slip bushings relieved?

Is it advisable to pilot the reamer in jig? If so, have bushings been provided?

Have arrangements been made to hold these bushings in place while piece is being drilled or reamed?

Is there anything to prevent soda-water or compound from getting on the work as it should?

Can the jig be kept from turning while the drill or reamer is in action? Can you use any of the standard handles?

Will the drilling come within the base of the jig at all times?

If the jig is a rotary jig, is the indexing positive and accurate? Can you use a straight index pin instead of a tapered one?

If the part can be tapped or spot faced in the jig to advantage, have you provided large enough slip bushings?

Has the counterbore been provided with stop collars?

Has the drill press the necessary speeds for drilling and reaming all holes? Must it have a tapping attachment also?

Always remember that it is not practical to have several small holes and only one large one to be drilled and reamed in the same jig, for the reason that quicker results can be obtained by drilling the small holes on a small drill press, while, if there is only one large one, it would require the jig to be used on a large machine. The question then arises, is it cheaper to drill the large hole in another jig and will the result, after so doing, be accurate enough?

Is the jig too heavy to handle?

Can a drilling angle, as shown, be used to advantage to take care of a hole which is on an angle?

Can the standard wrenches and handles be used with jig?

Are there any gages to be designed to help the operator get quick and accurate results from the jig?

Can you use a double or triple thread on the screw which holds the work in jig, so that it will take fewer turns to get screw out of the way in order to remove part more quickly?

Are there any loose parts of the jig, such as clamps, etc., that could be made integral with the jig and thus prevent their getting lost?

Have you made a note on drawing or have you stamped all loose parts with some symbol indicating the jig they were made for so that in case they are lost or misplaced they can be returned to the jig when found?

Are all necessary corners rounded?

Is there any danger of the operator being injured through the faulty design or makeup of the jig?

Can the toolmaker make the jig?

Are your drill bushings so long that it will be necessary to make up extension drills?

Are the legs on the jig long enough to allow the drill, reamer or pilot of reamer to pass through the part a reasonable distance without striking the table of the drill press?

Have you provided against clamps turning?

Are all clamps located in such a way as to resist or help resist the pressure of the drill?

Will a profile on base of jig help you to locate part to be drilled accurately?

Have you provided springs under clamps or bushings?

When tightening down on clamps, is the work apt to spring?

CHAPTER VII

JIG CLAMPS AND CLAMPING METHODS

The proper designing or making of clamps or clamping devices is one of the most important points in tool designing, and it too frequently happens that quantities of work are spoiled and needless injury often befalls the operator through their faulty construction.

There are but four things to be considered in the designing of a clamp; strength, the correct point for applying power on the clamp; the bearing points of the clamp; and the means provided to prevent clamp from turning when tightening down the nut or bolt.

Figs. 137 and 138 are three point bearing clamps, two bearing points in front and one in the rear. Attention may be called to these bearing points being rounded, which allows the clamp to properly seat itself on the work and to compensate for any variation in the size of the part to be clamped. This is particularly useful when a rough casting is to be operated on.

It is well to provide a spring underneath the clamps, as shown in Fig. 138, this will reduce the time otherwise required to manipulate clamps of this description.

Figs. 139, 140 and 141 are similar to Figs. 137 and 138, except the two points of bearing are in the rear and the single point in the front. Fig. 139 is the style of clamp frequently used for holding drill jigs, milling fixtures, lathes, fixtures, etc., to the table, bed or face plate of the machine.

Figs. 142 and 148 are also similar and are generally used in box-jig construction. They can be easily removed by sliding the plate longitudinally. The clamping screw is carried in a plate which is held in place by the retaining grooves.

Figs. 143 and 144 are good examples of what are generally called hook bolts; the bearing point of the hook should always be as close to the bolt as is possible.

Fig. 145 is a design of clamp not frequently seen in use, although for light clamping it has many advantages, it is quick to operate and takes up but very little room, which is so important in tool designing and making.

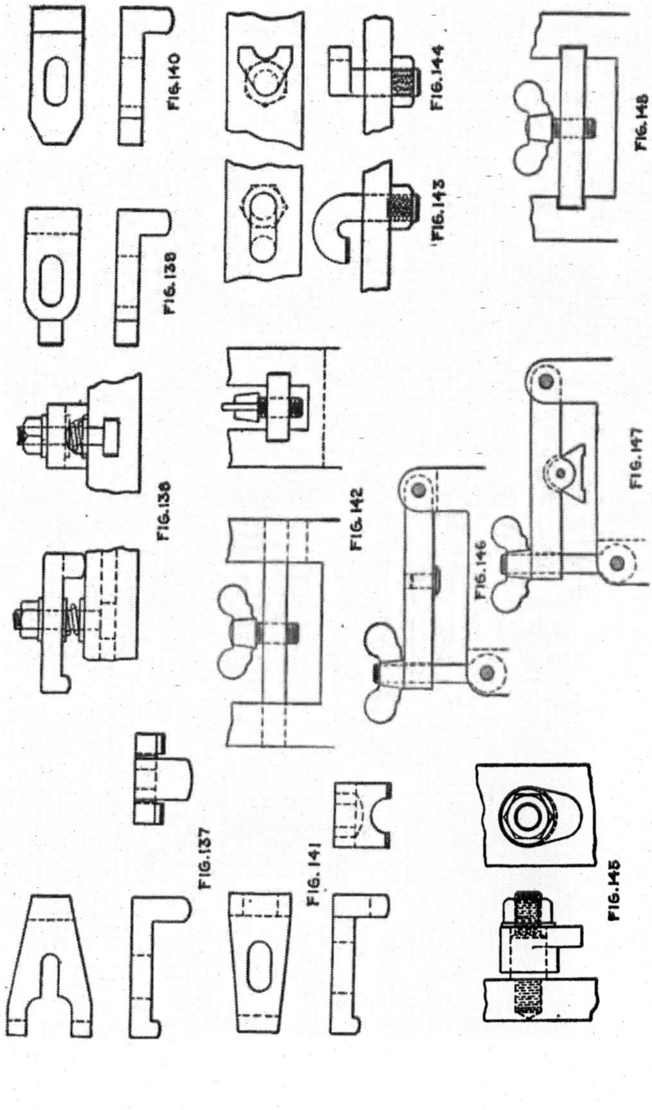

FIGS. 137 TO 148.—JIG CLAMPS AND CLAMPING METHODS.

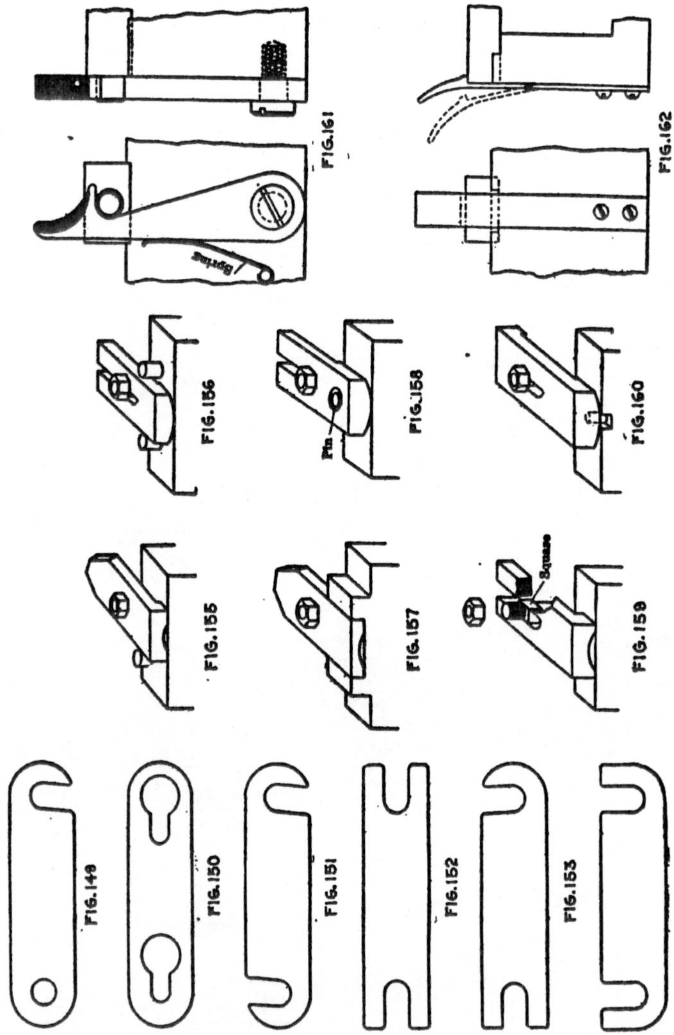

SEVERAL FORMS OF STRAPS AND LATCHES

FIGS. 149 TO 162.—SEVERAL FORMS OF STRAPS AND LATCHES.

Fig. 146 is the style of clamps most frequently used in drill-jig construction, especially table or box jigs. With this design the piece is held in the jig by one point striking in the center. In Fig. 147 the same style of clamp is used, except that it has two self-adjusting bearing points and in some instances three points.

Figs. 149 to 154, inclusive, show a few designs for jig straps which are the most commonly used. Fig. 149 is what may be called the swinging type; the others are all removable. The material

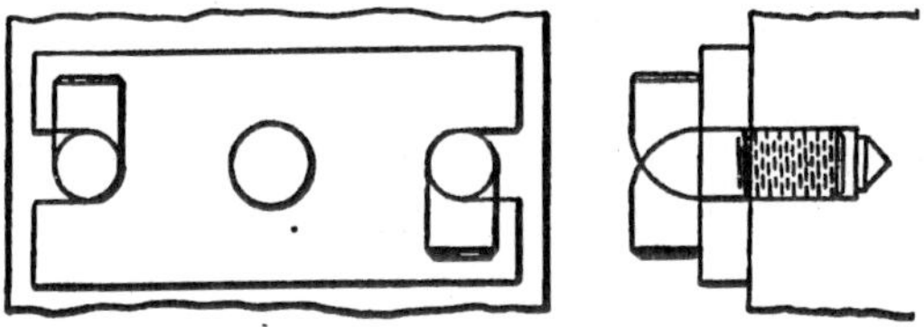

FIG. 163.

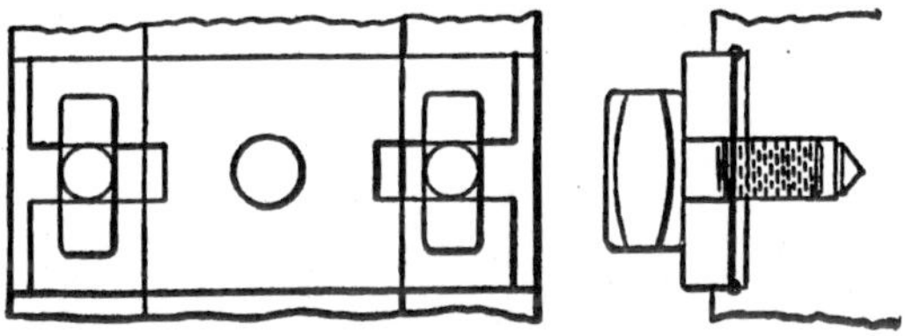

FIG. 164.

FIGS. 163 TO 164.—HOOK AND T-HEAD SCREWS.

generally used for making straps of this description is a good grade of machine steel.

Figs. 155 to 160, inclusive, illustrate the principal methods of holding clamps from turning while the nut or bolt is being tightened on them. The tool designer or maker too often overlooks this important point and does not seem to realize the advantage of properly holding them.

Figs. 155 to 157 are styles commonly used in drill-jig construction, while Figs. 158 to 160 are most frequently found in use on milling fixtures. The object of securing clamps in this manner is to get more rapid production from the tool to which they are assigned, and if only a second can be saved in the process of tightening down the clamp on the work, that second means many hours, if a great quantity of work is to be done.

Figs. 161 and 162 show two designs of latches, such as are used on small jigs for holding down a lid or plate which contains bushings or binding screws. Latches are made in many different styles. Sometimes the hook screw or lock screw is styled as a latch, probably because it is used for the same purpose. Examples of these screws are shown in Figs. 163 and 164. It is unnecessary to describe them further, as the illustrations speak for themselves.

Fig. 165 shows the design of an eccentric pin which is removable, but by threading the end and using a nut and washer it can be held in place, thus making it stationary, as indicated by dotted lines.

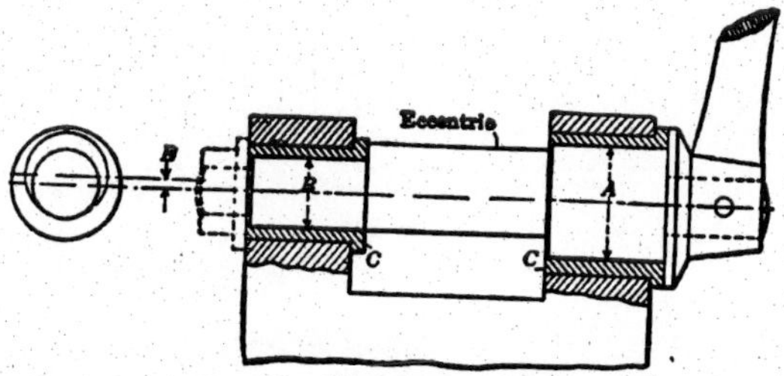

FIG. 165.—ECCENTRIC PIN LOCK.

The eccentric of *E* varies from 1/16 to ¼ inch, depending entirely upon the size of the pin. The diameters *A* and *B* are made a sliding or working fit in the bushings *C*. These bushings are generally made of cold drawn steel and case hardened.

Fig. 166 is another design of eccentric pin working in the body of the jig, and the illustration shows it operating two bolts *B B*. These bolts are in turn attached to a strap *A*. The pin, it will be noticed, is eccentric on both ends.

When the lever *C* is forced down the screw *E* will bear against the work and in case of any variation in the size of the work, such as a rough casting, this screw can be adjusted to suit.

The washer *D* should be rounded on the side bearing against the strap, which will allow an equal clamping stress on each of the bolts.

Figs. 167 and 168 show two designs of swinging binders, which are both used considerably in drill-jig construction. Fig. 167 is the most popular for the reason that it can be made very small and by a good proportion its strength maintained. It has the

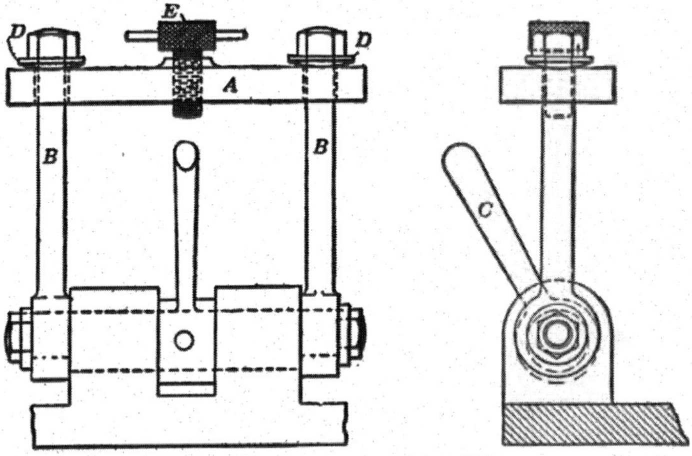

FIG. 166.—ANOTHER ECCENTRIC PIN.

advantage over Fig. 168 of having two screws instead of one to take up the thrust of the binding screw.

Figs. 169 to 172 show the construction of the sliding bushings which are so often used for holding the work as well as piloting the drill. The clamp is used in all cases for forcing the bushing against the work. Attention may be first called to Fig. 169 in this design, the bearing points *R* and *A*, are rounded to compensate for any variation in the hight of the work and always allow the clamp to properly seat itself preparatory to the thrust of the binding bolt or nut.

A spring *C* is placed directly beneath the clamp so that it will automatically bring the clamp back to its original position when the nut is released. If the slot in the clamp through which the bolt enters is elongated, it is advisable to place washers beneath the nut and between the clamp and spring as shown.

Figs. 170 to 172 are of similar design. The spring in this case, instead of being placed under the clamp, is directly beneath the head of the clamping bushing in Fig. 170; this spring is a coiled spring and is protected from dirt and chips by a special y designed head on the bushing. In Fig. 171 a flat spring is shown, in this case it is not necessary to protect the spring from dirt or chips, etc. In either case these two designs have the advantage, as the spring raises the bushing as well as the clamp as soon as the binding nut

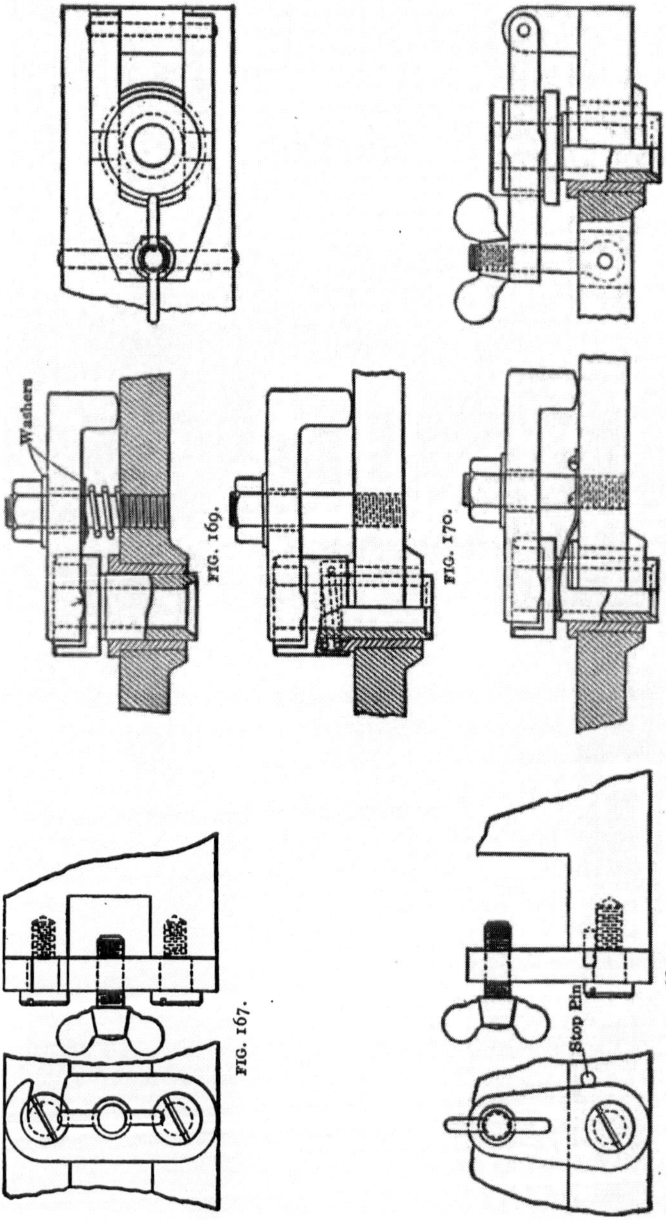

FIG. 167. FIG. 168. FIG. 169. FIG. 170. FIG. 171. FIG. 172.

FIGS. 167 TO 172.—SWINGING BINDERS AND SLIDING BUSHINGS.

is released, while in Fig. 169 the clamp is only released, and it becomes necessary for the operator to lift the bushing when removing or placing the part to be drilled.

Fig. 172 shows another way of operating these bushings. It is somewhat objectionable, owing to the fact that it is expensive and it has no advantages over the preceding designs.

SEVERAL GOOD CLAMPING DEVICES

One of the most difficult problems confronting the tool designer is the means for properly and securely holding the work. This difficulty most frequently arises when the part has no projection or lugs, so that it is impossible to use clamps.

The illustrations are shown to assist the tool designer or tool maker when in want of schemes for holding such work as plates, covers, etc., to be milled, profiled, disk-ground, etc.

A cheap, efficient and quickly-made work-holding device is shown in Fig. 173. The whole thing may be made on a lathe. The body *B* should be made of tool steel and properly hardened and ground. Care should be given the hardening as the sharp corner which is continually coming in contact with the hard, scaly casting, if too hard, breaks. The body is held in position by the nut *C* and washer *D* and should be a sliding fit in the support or body in which it works. A stiff spring should be inserted to act as a release. This scheme is excellent for holding light aluminum castings, and the construction prevents the chips from interfering with its movement.

A single point guided by the two gibs *A* and *B* and operated by the screw *C* is shown in Fig. 174. This screw controls the clamping and releasing movement of the nipping block *D* and does away with a spring which would otherwise have to be added to make the design correct. The screw should be placed as near to the center line of thrust as possible. This will reduce the clamping action on the nipping block which is the only fault in this design.

Both Figs. 175 and 176 are very much alike. Fig. 175 is frequently used when heavy work is to be done, and Fig. 176 when light milling or profiling is required.

The clamp shown in Fig. 177 is used when the part which it is to hold has to undergo a heavy cutting operation. When pressure is applied to the clamp, it not only has a tendency to force the part down but also ahead. An objection to this style of clamp arises when the designer makes too slight a taper at *B;* this causes the

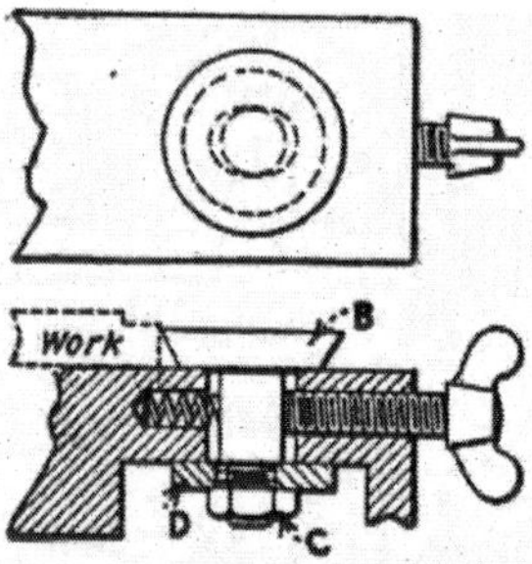

FIG. 173.

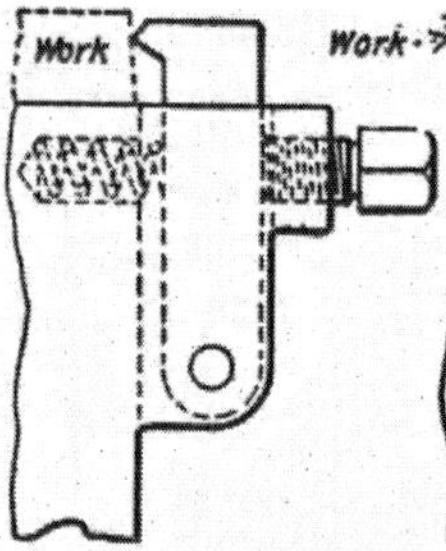

FIG. 175.

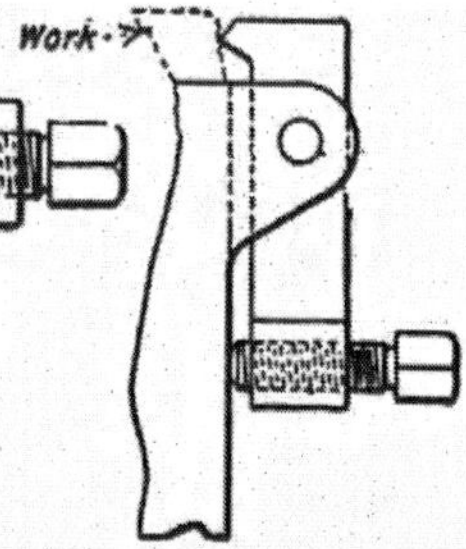

FIG. 176.

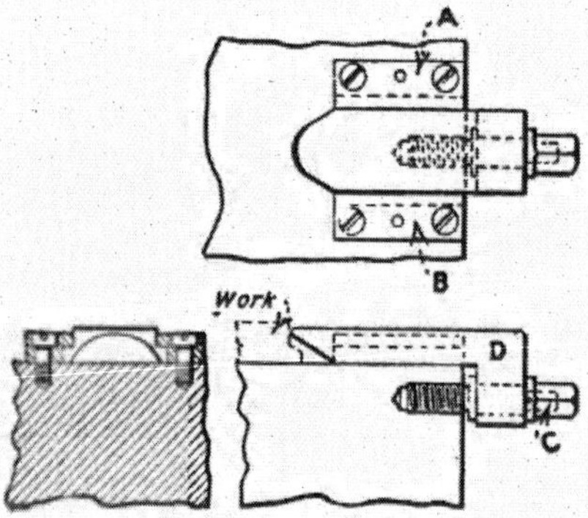

FIG. 174.

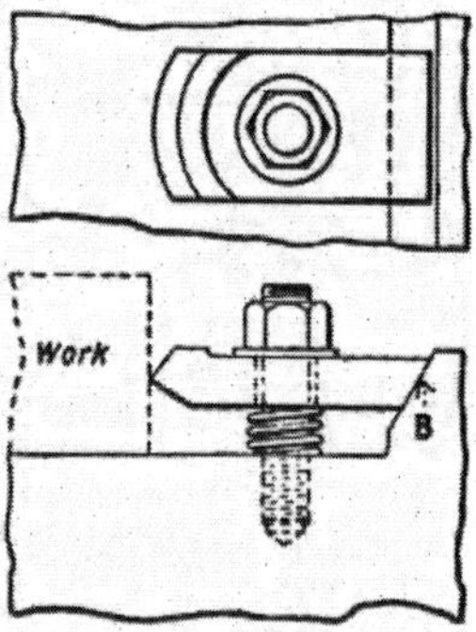

FIG. 177.

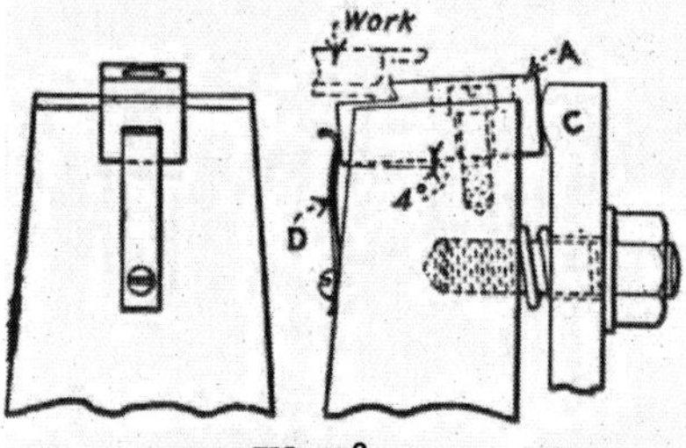

FIG. 178.

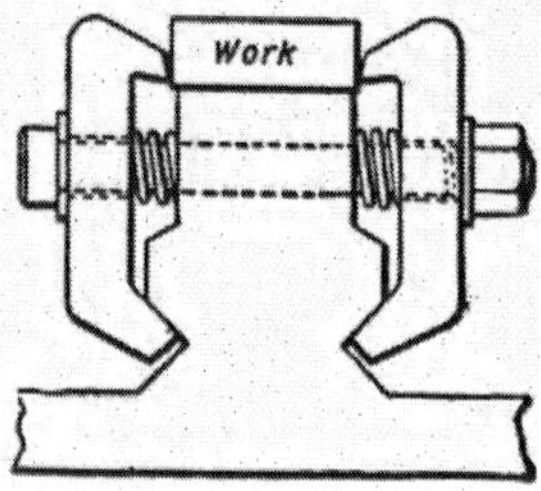

FIG. 180.

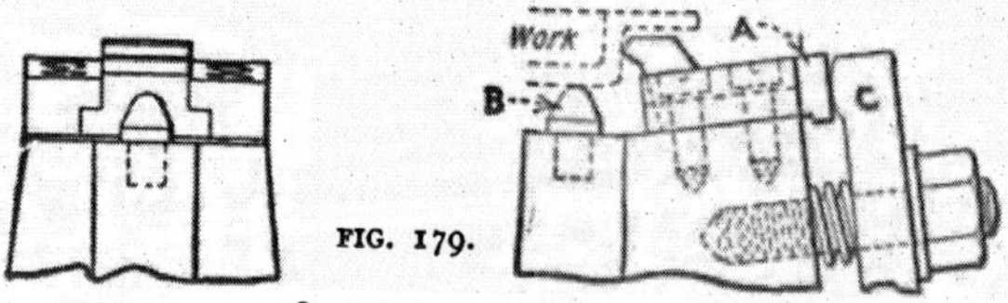

FIG. 179.

FIGS. 173 TO 180.—SEVERAL GOOD CLAMPING DEVICES.

clamp to bind between the work and taper. The taper should not be less than 30 deg. and a stiff spring should be added beneath the clamp.

OTHER ARRANGEMENTS OF NIPPING BLOCKS

There is a similarity between Figs. 178 and 179, the only difference being the nipping blocks *A*. In Fig. 178 the work rests on the nipping block *A*, which acts as a support for the work and, therefore, does away with the pin *B* in Fig. 179, on which the work rests in this figure. Both of these designs are used to good advantage on profiling or milling fixtures, especially when the work is large and of irregular shape such as aluminum crank-case basins, splash pans, gear-case covers, etc.

The nipping blocks should be a nice sliding fit and the pressure clamp *C* should be slightly rounded at the point of contact, so as to insure a perfect seat when the pressure is applied. Means should be provided to prevent the clamps from turning, so that when tightening the nuts upon them, they will stay in position. There should also be some provision similar to the flat spring *D* for forcing the nipping blocks back to their original position as shown at *D* in Fig. 178. There are many ways in which this can be done but the best method is generally governed by the design of the fixtures.

TWO CLAMPS AND ONE BOLT

The operation of two clamps with one bolt is shown in Fig. 180. The clamps not only hold the work like a vise but also nip and draw it down against the surface of the fixture. They should be

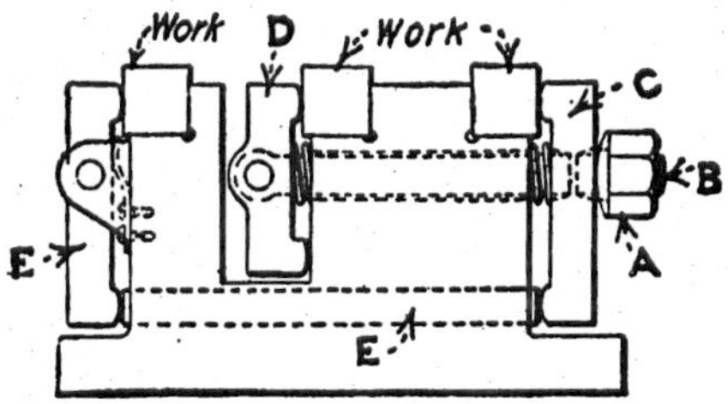

FIG. 181.—THREE CLAMPS ON ONE BOLT.

hardened, and suitable springs placed beneath them so that as soon as the pressure against them is released they will spread open and thus enable the operator to clean the surface of the fixture and also insert the part to be machined.

One of the important points in designing milling or profile fixtures is to do away with as many clamping bolts as possible. Operators waste much time tightening up several pressure clamps when one would be sufficient. These clamps are frequently in very awkward positions. For this reason Fig. 181 is shown to illustrate what can be done by the tightening of one bolt or nut.

As the nut *A* is tightened it draws the bolt *B* forward, thus tightening the clamp *D*, at the same time acting against the clamp *C*. The lower end of *C* also forces the rod *E* forward, and it in turn works against the clamp *F*. Thus the three clamps are operated by one nut only.

A QUICK-ACTING SCREW

A quick-acting screw is shown in Fig. 182. The handwheel shown suggests hand operation, but it can also be operated by the wrench if the handwheel is replaced by a nut. This of course depends upon the design of the fixture and the rigidity with which

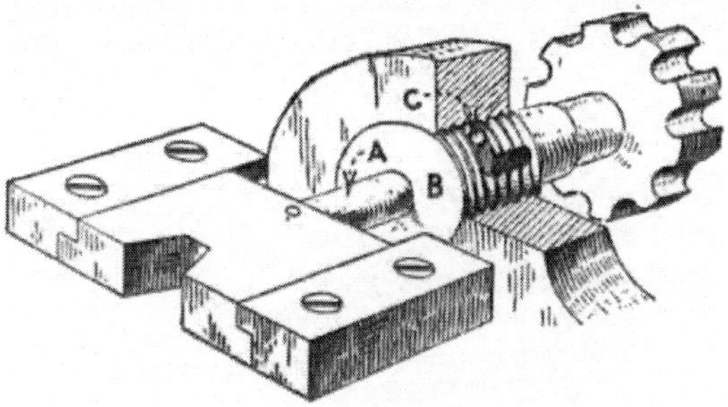

FIG. 182.—A QUICK-ACTING SCREW.

the part is to be held. This screw will be found useful in operating a V-block, as it overcomes the necessity of giving an ordinary screw several turns in order to get it back far enough to clear itself. The part *A* should be a sliding fit in the threaded collar *B*, which has a splined hole in which a quarter-inch pin *C* works. The length of this splined hole governs the movement or clearance wanted for operating the V-block. The dimensions given have worked out successfully for all classes of jig work.

An important point in making this scheme work out satisfactorily is to make the tapped hole in which the threaded collar works a rather stiff fit. The reason for this is that when the operator reverses the pin A, the collar will stay in its place, thus allowing the pin C to free itself from the locking end of the splined hole.

A rapid clamping device for jigs and fixtures is shown in Fig. 183. It can be used to the best advantage on drill jigs and fixtures

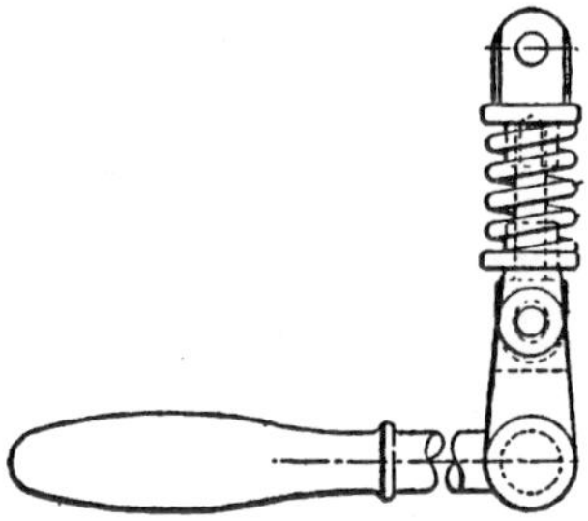

FIG. 183.—A RAPID CLAMP.

and is one of the quickest clamping devices known. It can be adapted to the operation of V-blocks like the quick-acting screw design and is brought against the edge of the piece. The engraving shows the device in its locked position. It will be noted that the spring is very heavy; this is necessary, as the scheme is to have the spring take up whatever thrust occurs on the part to be held, this thrust cannot, or rather should not, exceed the tension of the spring.

CLAMP BUSHINGS

Having seen such good and satisfactory results with clamp bushings, it is surprising that they are not more extensively used in machine-tool work. This probably is due chiefly to the difficulties that are encountered in their manufacture.

When in the employ of a large machine-tool maker recently, we persuaded the authorities to experiment with clamp bushings of the design shown in Fig. 184. The experiment being successful, the bushings were incorporated in a new design of machine, and the first order was for 1000 of four different sizes.

It was debated whether these could be stocked as standard parts, as the surfaces E, would have to be machined with the bore A. The holes A and B were both to be bored at one setting in a jig, and as we could rely on the dimension C being held, we decided to try and finish the clamp bushing right up; that is, by milling the surfaces E with a radius form cutter in a fixture made for this purpose. This proved much cheaper than boring them out with the hole A.

The results obtained were so satisfactory that it was decided to introduce a list of standard clamp bushings. They were tabulated

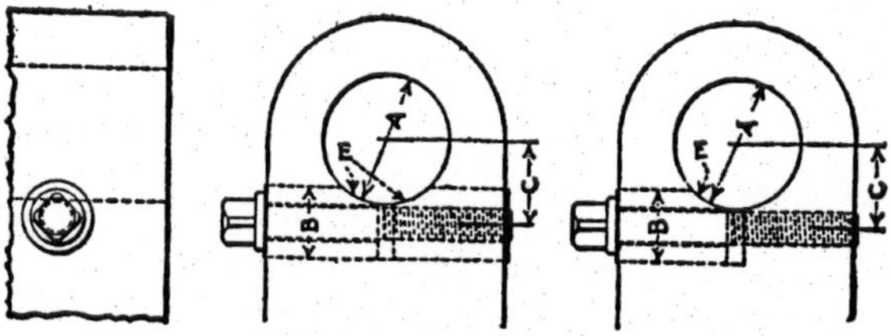

FIG. 184.—CLAMP BUSHINGS.

and made up in lots of 500 and put in stock. Both halves (in Model 2, Fig. 1, only the front half of the bushing is used, and the thread is tapped into the body of the piece) were made from ground stock,

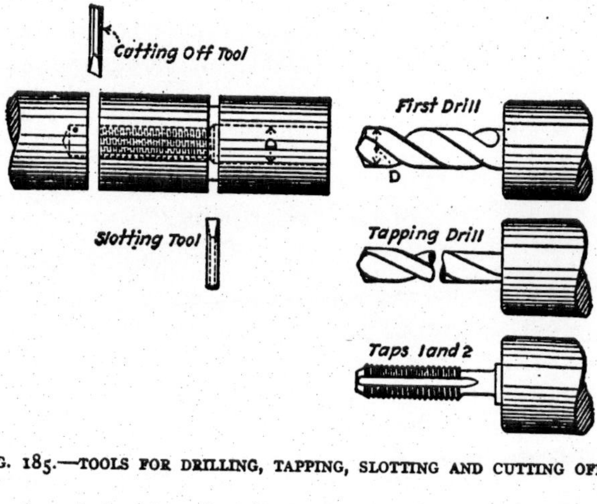

FIG. 185.—TOOLS FOR DRILLING, TAPPING, SLOTTING AND CUTTING OFF.

in one piece, being slotted, drilled, tapped and cut off on Cleveland automatics, as in Fig. 185.

The fixture for milling is seen in Fig. 186. Four pieces can be clamped and the cutter run straight through. The pieces are a nice fit in the holes A and are clamped by the bushings B. The body is made of cast iron and the bushing of soft machine steel.

The body is of rugged design with the tongue *C* to fit slots in the miller table; it is supplied with two edges *D*, to clamp the fixture to the table.

The long piece *E* with the tongue *F* is a nice fit on the dimension *H*. The tongue *F* is slightly thinner than the slot in the work piece

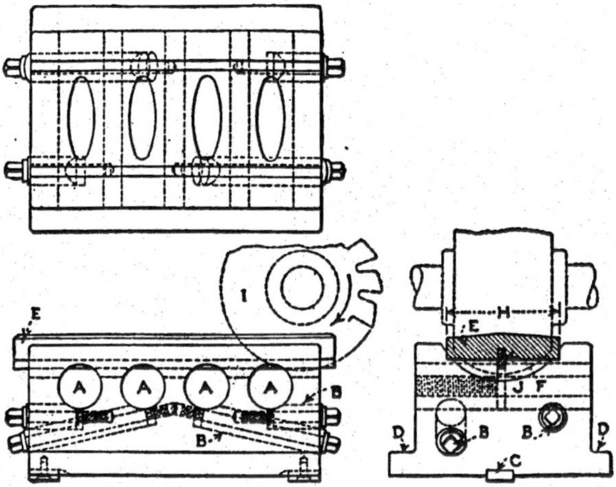

FIG. 186.—THE MILLING FIXTURE.

and lines up the pieces, insuring their centrality with the cutter. The piece *E* is removed before milling. The cutter is shown at *I* and the work piece at *J*.

One great feature of this fixture is, that as we are making clamp bushing of the same diameter of different lengths (*G* and *L*; see table of sizes), no change is necessary on the fixture for the different lengths.

The two halves were cut in two, and the corners rounded on a simple cutting-off machine. The bushings are not hardened, as we have found that they clamp better when soft and are less likely to injure the piece being clamped. They are made from ordinary cold-rolled steel.

While the dimensions in Table 19 may not be found to suit all classes of work, they have been found satisfactory in general machine-tool design. In making Model 2, two halves are made in one piece, as in Model 1, but the hole *D* is drilled through both halves.

Model One. Model Two.

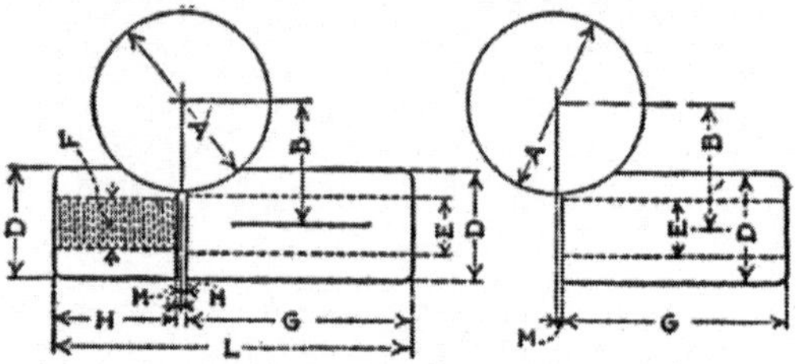

Diam. to be Clamped A	B	D	E	F	G	H	L	M
5/8	1/2	9/16	21/64	5/16–18	1/2 & 3/4	1/2	1 1/8 & 1 3/8	1/16
3/4	5/8	11/16	25/64	3/8–16	9/16 & 3/4	9/16	1 1/4 & 1 7/16	1/16
7/8	11/16	3/4	[illegible]	7/16–14	3/4 & 1 1/8	3/4	1 5/8 & 2	1/16
1	3/4	7/8	[illegible]	7/16–14	7/8 & 1 1/4	3/4	1 3/4 & 2 1/8	1/16
1 1/8	7/8	1	[illegible]	1/2–13	1 & 1 1/2	7/8	2 & 2 1/2	1/16
1 1/4	15/16	1 1/8	37/64	9/16–12	1 1/4 & 2	1	2 5/8 & 3 1/8	1/16
1 3/8	1 1/16	1 1/8	37/64	9/16–12	1 1/4 & 2	1	2 5/8 & 3 1/8	1/16
1 1/2	1 1/8	1 1/8	37/64	9/16–12	1 1/4 & 2	1	2 5/8 & 3 1/8	1/16
1 3/4	1 1/4	1 1/4	37/64	9/16–12	1 5/8 & 2 1/4	1 1/4	3 & 3 3/8	1/16
2	1 3/8	1 1/4	41/64	5/8–11	1 7/8 & 2 1/2	1 3/8	3 3/8 & 4	1/16
2 1/4	1 1/2	1 1/4	41/64	5/8–11	1 7/8 & 2 1/2	1 3/8	3 3/8 & 4	1/16
2 1/2	1 5/8	1 3/8	41/64	5/8–11	1 7/8 & 2 1/2	1 3/8	3 3/8 & 4	1/16
2 3/4	1 11/16	1 3/8	41/64	5/8–11	2 & 2 3/4	1 1/2	3 5/8 & 4 1/8	1/16
3	1 13/16	1 3/8	41/64	5/8–11	2 & 2 3/4	1 1/2	3 5/8 & 4 1/8	1/16
3 1/4	2 1/16	1 1/2	49/64	3/4–10	2 1/4 & 3	1 3/4	4 1/8 & 4 7/8	1/16
3 1/2	2 3/16	1 1/2	49/64	3/4–10	2 1/4 & 3	1 3/4	4 1/8 & 4 7/8	1/16
3 3/4	2 5/16	1 1/2	49/64	3/4–10	2 1/4 & 3	1 3/4	4 1/8 & 4 7/8	1/16
4	2 7/16	1 1/2	49/64	3/4–10	2 1/4 & 3	1 3/4	4 1/8 & 4 7/8	1/16

TABLE 19—SIZES OF CLAMP BUSHINGS.

CHAPTER VIII

MILLING FIXTURES OF VARIOUS TYPES

A milling fixture is in reality an extended and elaborate vise, and the designer of a fixture should therefore always first ascertain whether the job could not be as efficiently handled in a pair of special vise jaws. There are many milling fixtures which would never have existed if this fact had at first been fully considered. It should always be a distinct principle to utilize the existing equipment of the shop to its fullest capacity before launching out on an elaborate and expensive fixture by which nothing is gained.

Having arrived at the conclusion that the work is beyond the vise, we proceed to design a fixture. The next point is, how many of the given pieces will be required? If only a small number, the design should take quite a modified form, and the main thing to consider is to make the fixture as cheap as possible, but rigid enough to do the work and secure interchangeability. It is not necessary to pay any attention to the wearing surfaces, and the time required to clamp or change the pieces need hardly be considered.

When the quantity required is large, the problem takes quite another form, and the greatest consideration should be given to wearing surfaces and quick action.

The next point to be settled is the machine on which the job is to be done. It often happens that the machine most suited for doing that particular work has already more than its share of jobs allotted to it, and the proper authority should certainly be consulted on this point. It is hardly good shop economy to have one or two machines crowded with work, while others are standing idle because the fixtures are not adapted to them.

All jobs which can be machined by means of a butt or face mill, should be handled in this way in preference to using a cutter on the arbor of the machine. The gain in solidity and power is so considerable as to be almost self-evident. The recent increase in the number of vertical millers, some of which are also provided with side cutters, is an indication of the manufacturer's appreciation of this point. In these machines only butt and face mills can be used.

In straddle and gang milling where the cutters are on the arbor of the machine, the fixture should be so designed that cutters of the smallest diameter possible may be used. Nothing is so bad as a fixture with a projecting clamp on the top, which necessitates the

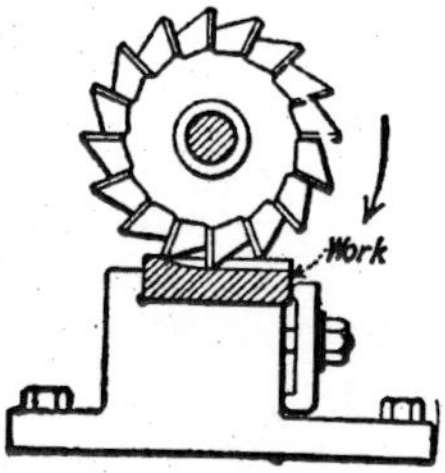

FIG. 187.

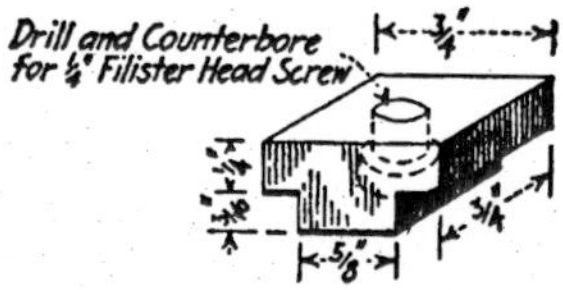

FIG. 189.

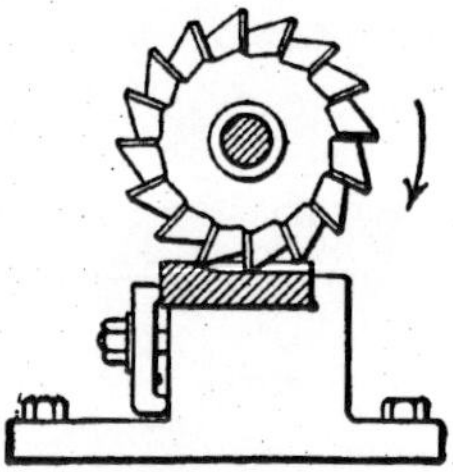

FIG. 188.

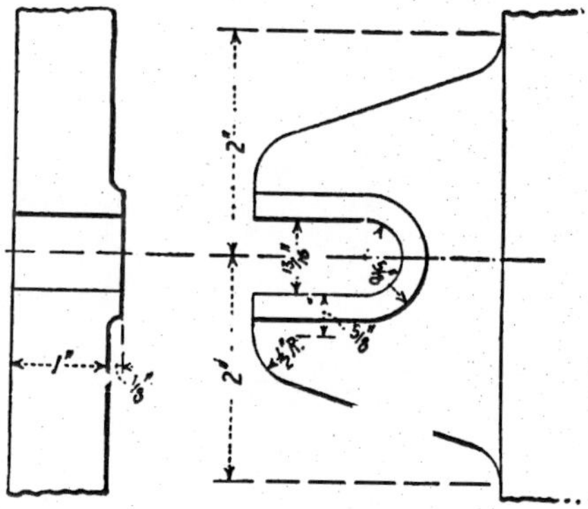

FIG. 190.

FIGS. 187 TO 190.—HOLDING WORK TO BE MILLED.

use of a cutter several inches larger in diameter than should actually be necessary.

To arrange the fixture so that the milling cutter shall be continually operating on one piece while another finished piece is being replaced, and thus continuous milling secured, should be the aim of the designer in all possible cases. Some manufacturers have recently made a great talking point of the adaptability of their millers in this way. The output of a machine may often be more than duplicated when this fact is fully realized and acted upon.

An important point, well known to tool designers, although sometimes ignored, is illustrated in Figs. 187 and 188. The work should always be secured in the fixture so that the thrust of the cutter can be taken up by a solid stop, as in Fig. 187, and not as in Fig. 188 where all the thrust is taken by the clamp and stud which are sure to spring more or less, and probably work loose through vibration.

TONGUES AND LUGS

A very good way of making a tongue-piece or key for locating the fixture in the slots of the miller is the table illustrated in Fig. 189. With this type of tongue a fixture is equally adapted to go on a machine with a 5/8- or 3/4-in. slot. The key is let into a square in the base of the fixture. By removing the retaining screw it may be turned either to allow the tongue to fit lengthwise in a 5/8-in. slot, or crosswise in a 3/4-in. slot.

It is often found that the slotted projecting lugs for attaching fixtures to the table of a machine are too weak for their purpose, and break off when in use. Fig. 190 gives the dimensions of a lug found sufficiently strong in practice, and these figures may be of some assistance to designers.

APPLICATION OF GAGES

It is often advantageous to use gages in connection with fixtures which either form part of, or are detachable from, them. Fig. 191 shows what is generally classed as a target gage. It is located on the fixture by the hardened pins *A* which fit the bushes *B* in the gage. The work is then adjusted until the center and small bosses correspond to their respective duplicates on the gage. After the piece is securely clamped, the gage is removed and the piece may be milled on its locating points as well as on the two sides.

The milling fixture shown in Fig. 192 has a fixed gage *A* by means of which the cutter may be set to mill the slot in the ring, which must be directly in line with the center. When the cutter has passed through the gage *A*, the table is lowered, moved forward, and raised until the cutter is in the correct position to mill the slot to the required depth. The slot is only cut on one side of the ring, and the gage is quite clear of the cutter after the job is once set.

A case where the gage is hinged to the fixture, and serves two purposes is illustrated in Fig. 193. The part *B* acts as a gage for the workman to set his milling cutters to, and the part *A* acts as a gage for properly locating the work. After the work is located in its correct position, and the clamp is firmly secured, the gage may be swung out of the path of the cutter, into the position indicated by the dotted lines.

In making the gage care should be given to the hardening, and all contact points should be ground. The illustration shows an operation, consisting of milling a slot and two sides, which can be successfully accomplished with this style of fixture and gage.

In Fig. 194 is shown the application of a fixed stop for setting

FIG. 191.

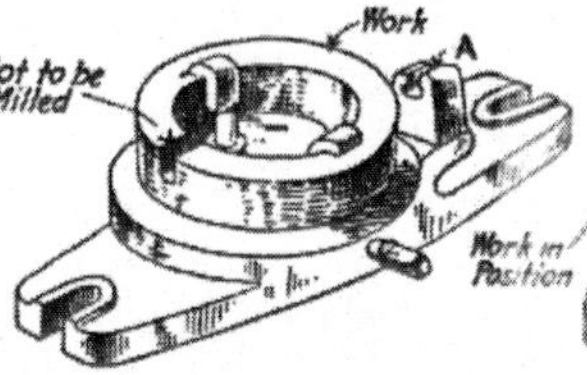

FIG. 192.

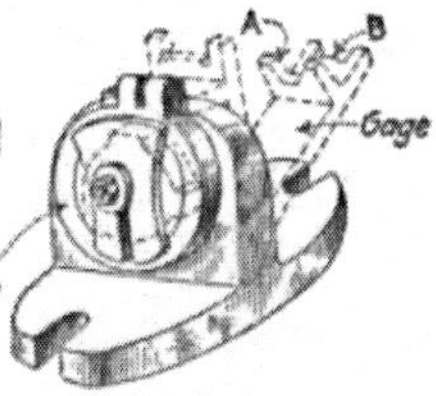

FIG. 193.

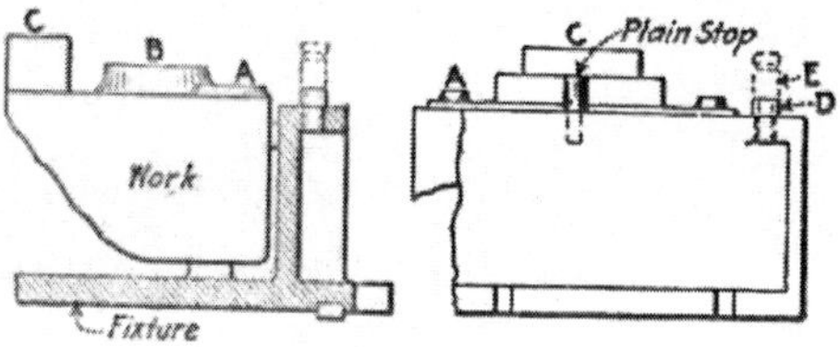

FIG. 195.

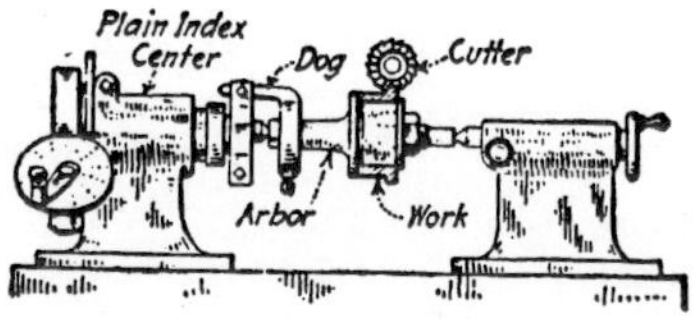

FIG. 196.

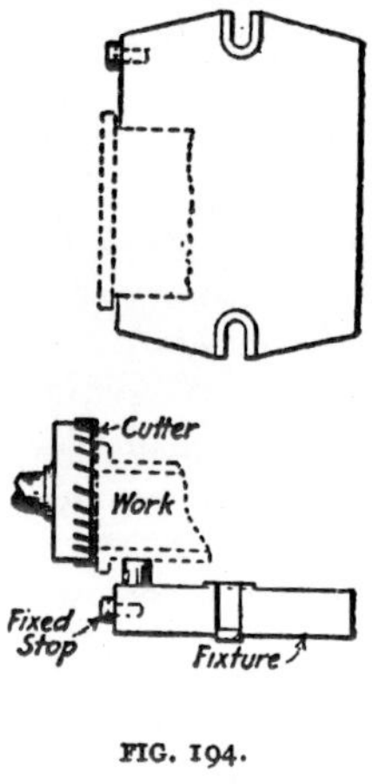

FIG. 194.

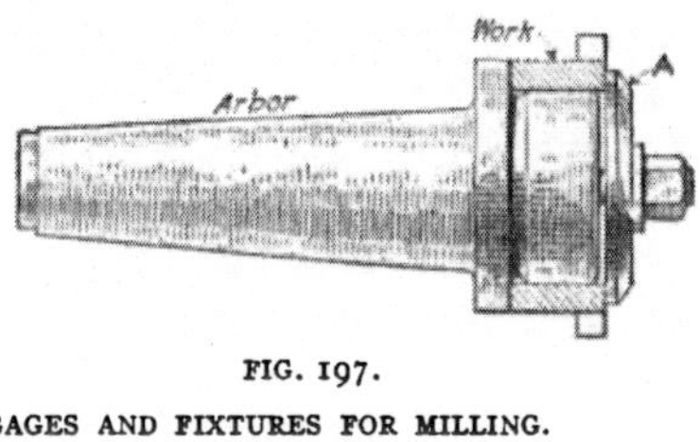

FIG. 197.

FIGS. 191 TO 197.—GAGES AND FIXTURES FOR MILLING.

the cutter to the correct height, while Fig. 195 illustrates the use of gages attached to the fixture for gaging multiple-height surfaces. For the surface *A* the fixed stop is used, which in this case is in the form of a bush *D*. For the surface *B* a plain stop is provided, while for the face *C* the knurled plug *E* is inserted in the bush *D*. By a combination of bushes of varying heights and a plug, any number of surfaces can be provided for without the stops in any way impeding the movement of the cutter.

INDEXING FIXTURES

A large amount of the work which the miller is often called upon to perform falls under the heading of "indexing."

Most of the universal millers in use are equipped with dividing heads, by which almost any division may be secured at any angle, but their use is more confined to the toolroom or experimental departments. In the factory a plain dividing head is more often used.

A general method of using the dividing index, in which the work is mounted on an arbor between centers for the purpose of milling slots at equal distances around the circumference is shown in Fig. 196. This is very elementary practice and should not be used for regular manufacture. There are too many connections between the indexing mechanism and the cutter, and the work is held in an excellent position in the middle of the mandrel, to enable it to spring away from the cutter; it also takes considerable time to change a part, as it is necessary to take the mandrel out of the centers.

A simple mandrel, Fig. 197, with a taper to fit in the spindle of the dividing head, forms a much more solid and rapid arrangement. After a half turn of the nut the slotted washer *A* can be removed and the work slipped on or off.

A substantial dividing head is illustrated in Fig. 198. There are many of these heads on the market, all of rigid construction and excellent workmanship. The handwheel forms a very ready way of rapidly indexing, and by designing suitable mandrels as shown in Fig. 197 and collets which are fitted to the taper hole in the head, quite a number of jobs may be handled.

A special dividing rig used for milling the slots in a piece of fairly large diameter is represented in Fig. 199. The work is rigidly supported behind to resist a heavy cut. It can be indexed very quickly, the handwheel and indexing pin being situated in a very convenient position. The work is held on the fixture by a nut and slotted washer.

A very rigid and durable fixture is seen in Fig. 200, which, owing to its form, has been nicknamed the "mud turtle." The index plate being underneath, is well protected from chips. It is an advantage to provide this fixture with a separate index plate for each number of divisions required. This is a distinct advantage over using a plate, say of 24 divisions, for general application. The work can be much more rapidly indexed when the operator is certain that the

FIG. 198.

FIG. 199.

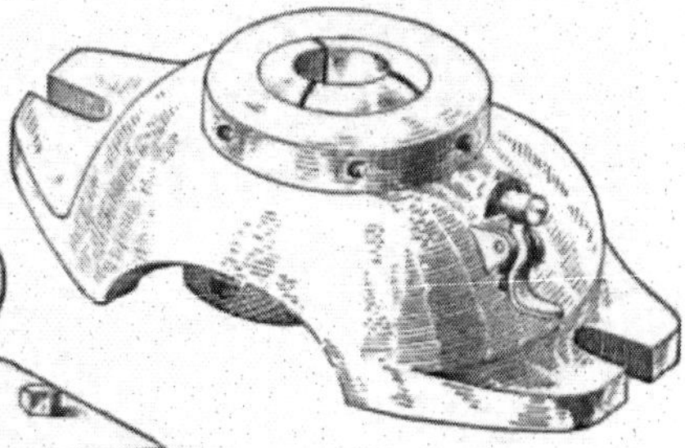

FIG. 200.

FIGS. 198 TO 200.—A MUD TURTLE SHAPED FIXTURE.

next hole the plunger drops into is the one required. Draw-in collets, similar to that shown, also form part of the equipment of this fixture. They are made in sizes to take stock, advancing in eighths, from 1 in. up to 2½ ins. and are of case-hardened and ground machine steel.

In Fig. 201 is shown an automatic indexing fixture used on a hand mill, for milling the small spanner slots in an adjustable ball bearing. The work is held by means of the expanding chuck in the center, operated by a square-headed, detachable key. The automatic indexing is secured as follows: The work having been fed against the cutter and one slot milled, the table is withdrawn until the pin *A* comes against the stop *B*, which is a fixture on the knee of the machine. When the table is still further withdrawn the locking lever C is pushed out and as soon as it has moved clear of the index plate, the pawl D engages in the ratchet and turns the work until the pin A comes in contact with the adjustable stop E.

The index plate should then be in a position for the locking lever to enter the next slot. As the table is again moved toward the cutter,

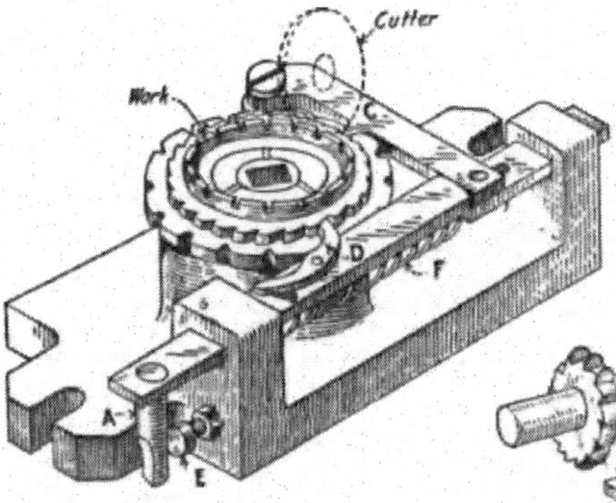

FIG. 201.

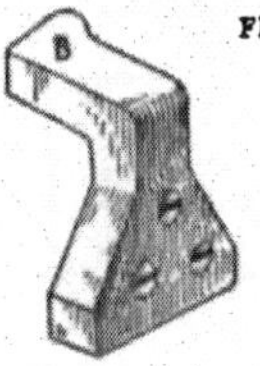

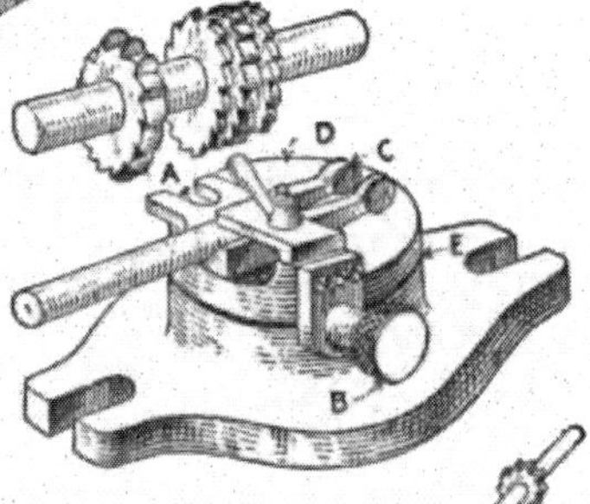

FIG. 203.

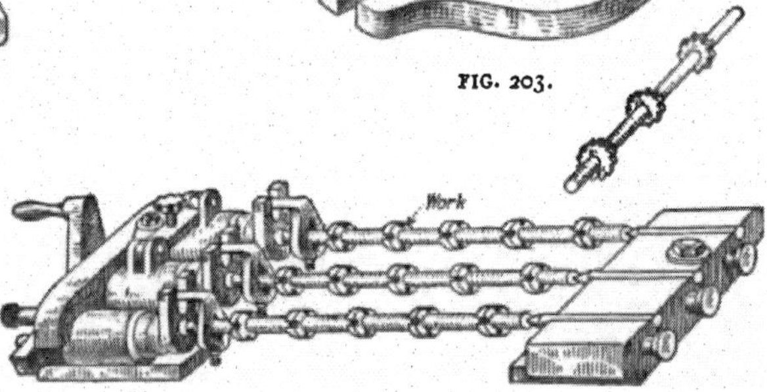

FIG. 202.

FIGS. 201 TO 203.—SINGLE AND MULTIPLE INDEXING FIXTURES.

the locking lever is forced into its slot by the spring F, and the work is in position for the next cut. In the illustration all the parts are shown exposed, but when working, the whole of the indexing mechanism is protected from chips and dirt by a sheet-steel casing.

A multiple indexing head and tailstock are illustrated in Fig. 202. This shows three shafts which can be simultaneously indexed for milling the four slots in the collars. The heads are rotated by gears keyed on the splindles, and engaging one with another. A handle is attached to the middle gear, which on being rotated turns the outside heads in the opposite direction. The direction of rotation being immaterial for indexing purposes, it is not advisable to complicate the fixture by using an idler. Care should be taken that no backlash occurs between the gears.

The fixture illustrated in Fig. 203 is an example of eccentric indexing. With this style of fixture two operations may be performed on a piece at one setting. The piece shown in the cut is in position for milling for rounded slot *A*. By withdrawing the plunger *B* the top plate may be swung around 90 deg. and again locked in place; the work is then opposite the gang of cutters for milling and facing the clevis *C*.

It should be observed that the whole of the work, where cutting takes place, is well inside the top plate *D*, and so there is no tendency to tip, the thrust of the cut being taken squarely on the circular bearing-ring *E*. This principle is capable of fairly wide application, as parts can be indexed at any angle, which brings them into correct position for the cutters.

The cam principle applied to the clamping of pieces in the milling fixture is illustrated in Fig. 204. The work is located by means of the pins *A* and held in position by the clamp *B*, which, in the sketch, is shown in the open position. This is locked in place by passing the cam *C* around the pin *D*. This clamping device should be used only in cases of very light milling, and cannot be relied upon for heavy duty.

Another type of cam is illustrated in Fig. 205, but its use is rather restricted owing to its small range of movement. Suitable proportions for these cams were given in a previous chapter on clamping devices.

The use of the wedge is shown in Fig. 206, which illustrates a method of holding a piece of basin-like form while milling the flanged joint. The material of the piece is aluminum, and to prevent springing and distortion of the part, the clamp is made to conform to the inside shape. It is locked in position by the wedge *A*. This plain wedge method is also found unreliable for heavy work, and its range of adjustment is limited. Pieces clamped by this means would be required to run somewhat uniformly.

In Fig. 207 is illustrated a type of fixture suitable for profiling the faces of irregular, pan-shaped pieces, where it is only possible to locate and hold them on the flanges. The flange of the piece rests on the three fixed uprights, and is held in position by the sliding guided clamps. The work is additionally supported by the adjustable jack pins, which prevent it from springing under the pressure of the cut. There are many ways of designing these adjustable jack pins.

An adjustable angle plate is shown in Fig. 208. This may be used in conjunction with the standard dividing head, miller vise, or special fixture, for covering a range of work milled at an angle, sometimes difficult to deal with in any other way. It should be used only for very light jobs, as the braces at the back are likely to move out of position under a too heavy cut. These braces should be replaced by a solid web, in cases where a large number of parts are to be machined at the same angle, sufficient to justify the expense of a special casting. The slot *A* should be made the same size as the miller table on which it is to be used; this will generally be either 5/8 or 3/4 in.

RADIUS AND FORM MILLING

Cases often occur in which it is necessary to mill an arc of large radius on a piece which cannot be handled by a form cutter, or which it is not considered desirable to machine by that means. The fixture shown in Fig. 209 is used for milling the arc on the end of the work, which is seen clamped in position. The necessary rotary motion is secured by rocking the movable portion of the fixture by hand, by means of the projecting handle. The handle should be quite long; this gives the operator sufficient leverage to resist the tendency to chatter. Small work only can be handled in this way.

The fixture in Fig. 210 performs a similar operation, but provides a much more substantial means for taking a heavy cut, although it is by no means so rapid. The rotary motion is obtained by turning the balanced handle *A*, operating the worm spindle *B*, which engages in the rear part of the plate *C*, rotating it about the fulcrum pin *D*. The plate *C* should be secured to the base in such a manner as to allow the movement only in the plane desired. There is a tendency to chatter if this is not provided for.

A profiling method adaptable for use on a small hand miller is illustrated in Fig. 211. The work shown is a square-ended shaft,

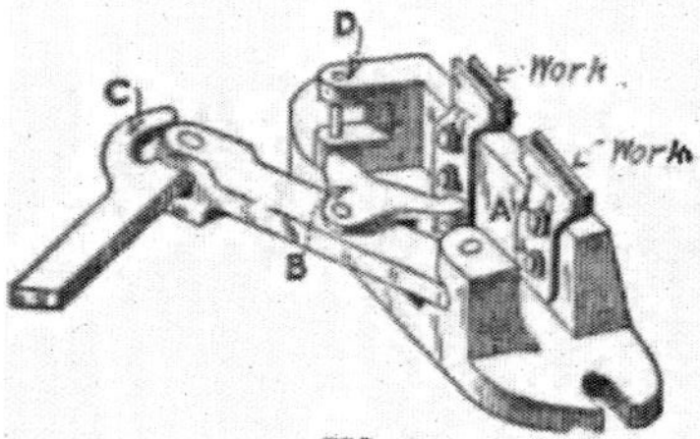

FIG. 204.

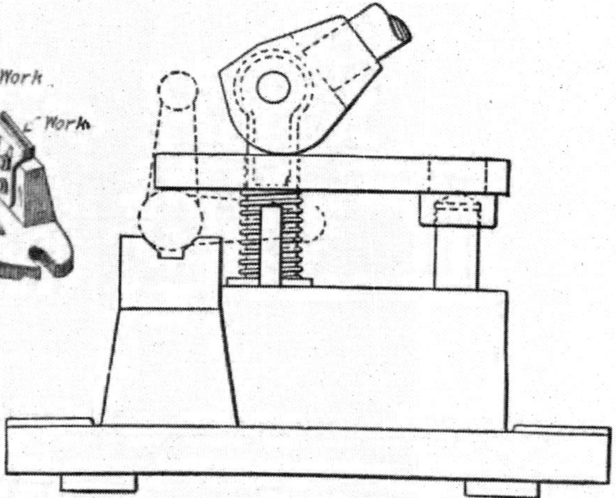

FIG. 205.

FIG. 207.

FIG. 208.

FIG. 209.

FIGS. 204 TO 209.—MILLING FIXTURES WITH CAMS AND WEDGES.

the sides of which are milled to a radius to enable it to rock; it forms part of a flexible coupling. At *H* is shown a section through the

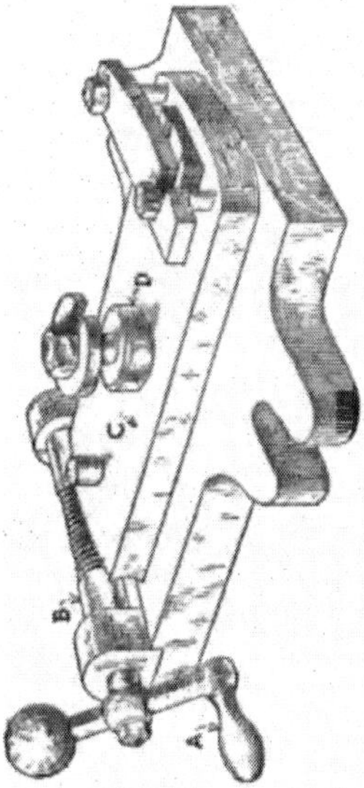

FIG. 210.

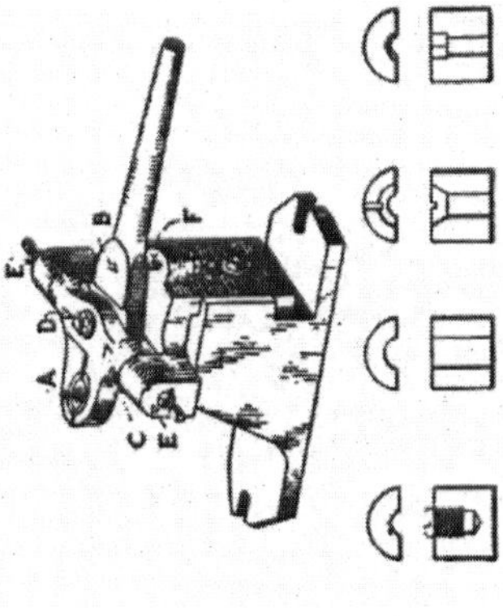

FIG. 212.

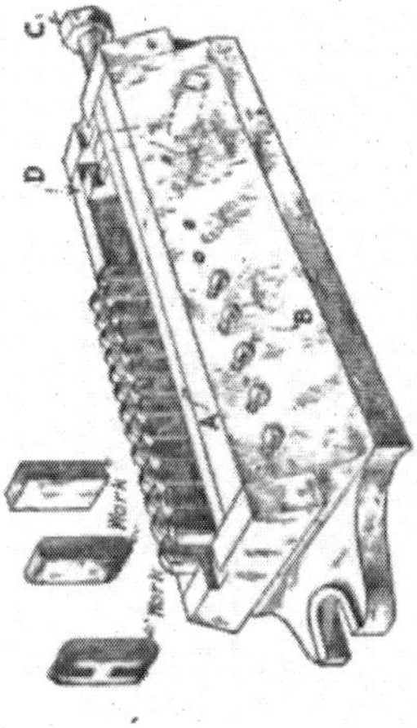

FIG. 213.

FIGS. 210 TO 213.—FIXTURES FOR RADIUS AND FORM MILLING.

roller and former, in which it can be seen that the roller is conical and that the former is at an angle to correspond to it.

By moving the table away from the machine the cutter can be brought down further into the work, until the required size is secured. An angle on the former of about 7 deg. gives a very convenient range of feed. The roller is made of tool steel, hardened and tempered, while the former is of machine steel case-hardened. The former can easily be taken off and others attached to the same centers of various lengths and forms to suit the part being machined.

A hand device suitable for milling the slots in small screws is shown in Fig. 212. The lever *B* is free to move in a horizontal direction about the point *D*, and the work is nipped between it and the jaw *C*. These two parts may be turned on the adjustable centers *E*, and the work raised against the cutter until lever *C* comes in contact with the adjustable stop *F*, when the slot should have been cut to its required depth. The whole upper part of the fixture is capable of being adjusted for height, and locked in position. The hardened bushings seen below are part of the equipment, to suit the various forms and sizes of the screws requiring to be slotted.

In Fig. 213 appears an example of the type of fixture suitable for holding a number of pieces at once, when the stock operated on may be relied upon to come uniform in size. The piece shown in the illustration is cut from rectangular, cold-rolled steel and the variation in size does not amount to more than a few thousandths. The strip *A*, which is supported by the adjusting screws *B*, is capable of taking care of any such slight differences.

The pieces are clamped longitudinally by the hexagon-headed clamp-screw *C* pressing against the hinged plate *D*. This has the effect of forcing the work down into the fixture when tightening up. The operation of rounding the ends of the work may also be performed in the same fixture by providing an additional packing plate to make up for the difference in the width of the part when turned in the other direction.

A fixture for holding a number of rough forgings is shown in Fig. 214. The hinged spacing blocks insure that each piece is held squarely, and compensate for any difference in the size of the piece; they also hold them down into the Vs around the small bosses. The double row takes full advantage of a powerful machine. There is a considerable waste of table travel between pieces owing to the necessity of having a space. The advantages of using a fixture of this kind lie mainly in the fact that one man can then attend to several machines; unless this is the case, it would probably be quite as quick or even quicker to use a simpler fixture taking only one or two pieces at a time.

FIG. 214.

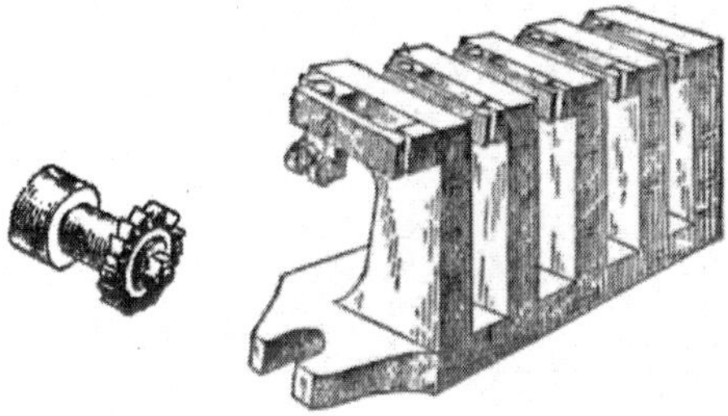

FIG. 215.

FIGS. 214 TO 215.—SUPPORTED AND OVERHUNG FIXTURES.

OVERHUNG WORK

A fixture in which the work is held above the cutter is shown in Fig. 215. The advantage claimed for this construction is that all the locating and clamping arrangements are kept free from chips. The work supports must be massive as there is necessarily considerable overhang. Unless the work particularly lends itself to easy locating and clamping in this type of fixture, it is questionable whether the advantage of having clean locating surfaces balances the incidental disadvantages of having the work and cutter hid from the view of the operator when he stands on the side of the machine where all the controlling levers are situated. It is also in most cases necessary to hold the work in place while clamping it in position. In the illustration it will be observed that considerable travel of the table is wasted in going from one piece to another; this cannot be avoided unless the rigidity of the fixture is sacrificed.

CONTINUOUS MILLING

A case in which continuous milling is partially secured appears in Fig. 216. The design is not ideal, but illustrates satisfactorily this style of milling. While the cutter is operating on the piece *C* the piece *A* may be changed; when *D* is being machined, *B* may be replaced and *A* clamped tightly down. The table is then lowered, run back, and raised to the correct height and the cut started once more, and while the cutter is at work on *A* the pieces *C* and *D* are put in place and the remainder of the clamps secured.

A similar principle is illustrated in Fig. 217. In this case, however, each piece is clamped independently, and a butt mill is used. It should be noticed that with a fixture of this kind a machine may be run at its highest feed and cutting speed. The work is right down on the bed of the machine and the overhang of the knee and cutter is as short as it is possible to get it.

Vertical milling in many cases presents several advantages over the horizontal style, as full use may be made of the reverse, traverse and crossfeeds without loosening that weak point in the miller—the knee joint.

A fixture, or rather two fixtures, by which continuous milling is secured, are shown in Fig. 218. The piece having been clamped in position in the fixture *A*, the cutter commences to operate on it, starting from the position between the two fixtures. While this is taking place another piece is put in *B*. The cutter having passed over the piece in *A*, it is reversed, the feed changed to a higher table travel, and the cutter passes back over the same part, taking a light skimming cut due to the piece having been slightly sprung in the roughing operation. The feed is then dropped to its previous rate, and the roughing cut commences on the piece in *B*; meanwhile, the piece in *A* is changed.

This method could be improved upon if the table were wide enough to take four fixtures, and the crossfeed utilized. The path of the cutter would then be in the form of a rectangle instead of a straight line and the finishing cut could, if desired, be dispensed with.

Gang milling the forked joint *X* is accomplished in the fixture shown in Fig. 219. The clamping arrangement is duplicated on a circular base which can be located on the rotary table seen in Fig. 220. The central boss *A* serves as a register and the nut *B* projects

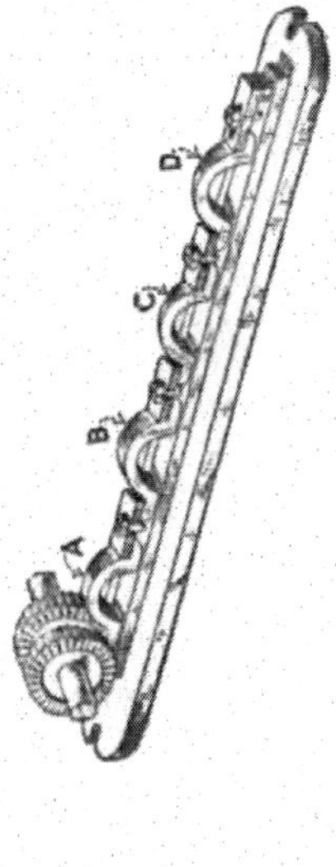

FIG. 216.

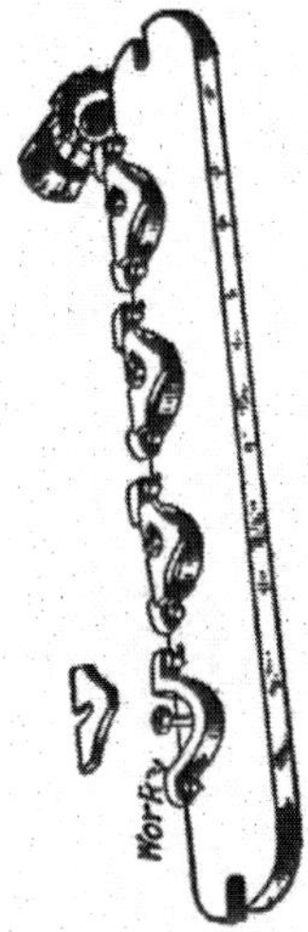

FIG. 217.

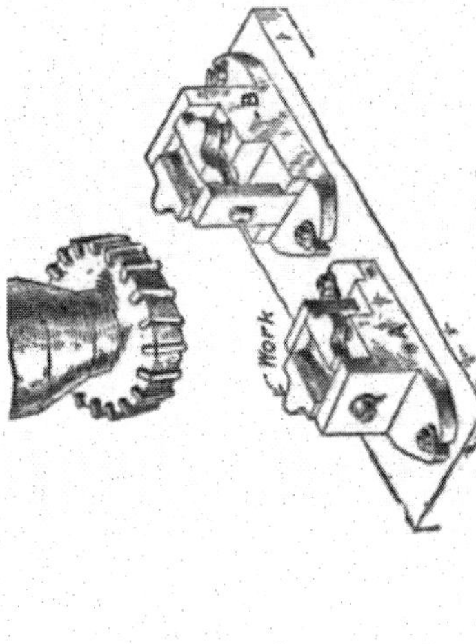

FIG. 218.

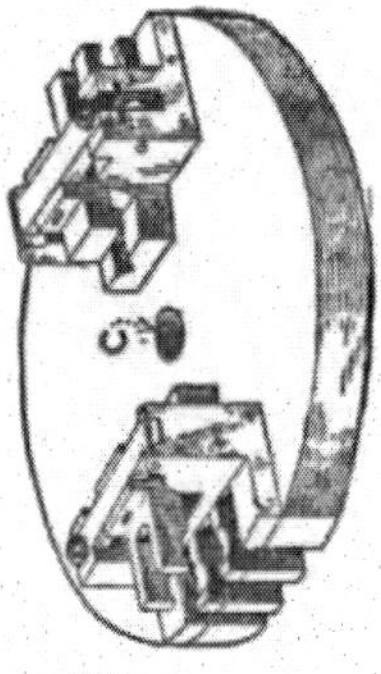

FIG. 219.

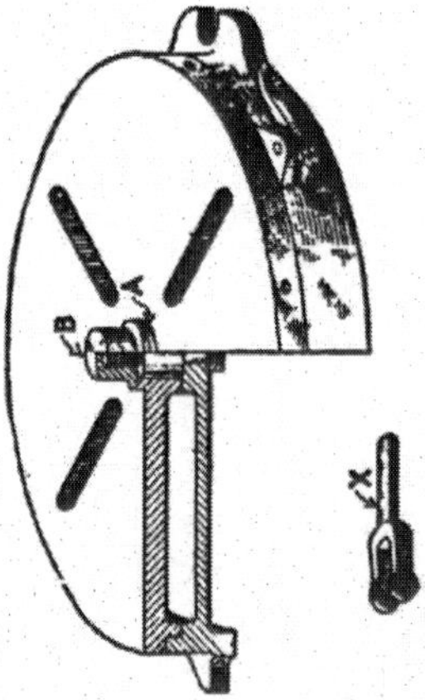

FIG. 220.

FIGS. 216 TO 220.—FIXTURES FOR STRING AND CONTINUOUS MILLING.

through the hole. This nut is tightly secured by a box spanner which is not shown. The diameter of the table is about 14 ins.

This fixture is used on a Lincoln miller, and while the cutter is

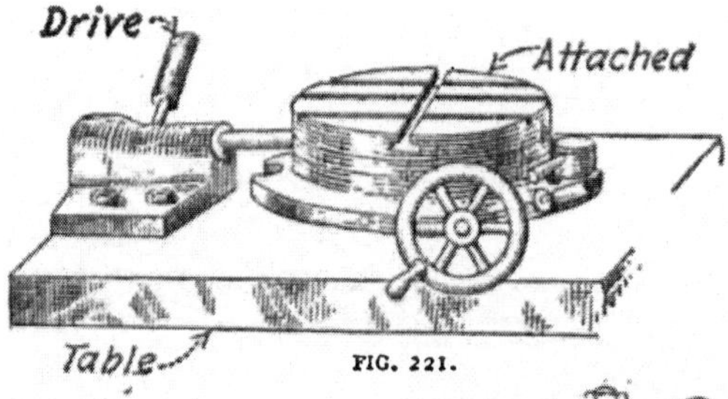

FIG. 221.

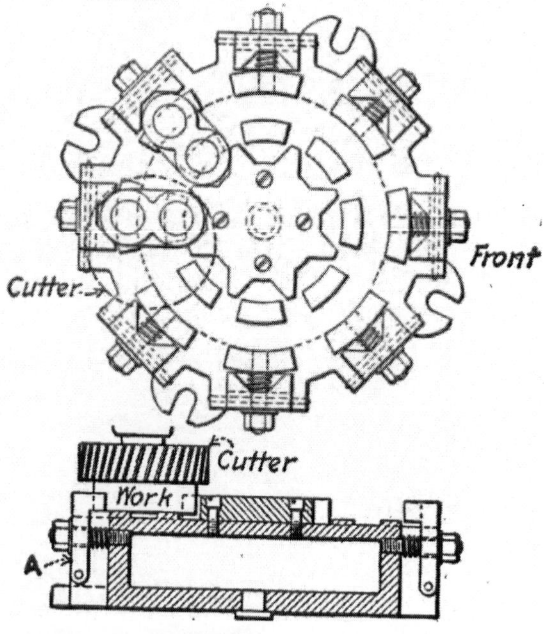

FIG. 222.

FIGS. 221 TO 222 —ROTARY FIXTURES FOR MILLING.

operating on one part another piece is being secured in position. As soon as one piece is finished the machine table is run back, the fixture swung around, the feed dropped in and the operator proceeds to change one piece while the machine is operating on the other.

A type of standard rotary attachment is shown in Fig. 221. It is provided with power or hand feed, and may be used on the table of any kind of a miller. A great variety of fixtures may be designed to go on this attachment, by which continuous work may be performed.

A rotary fixture, such as is used on the rotary attachment shown in Fig. 221 appears in Fig. 222. When used on a vertical miller the nearest approach possible to continuous milling is attained. In the illustration provision is made for holding eight pieces, which are secured independently from the side. The clamp *A* is hinged, and when the nut is loosened it is pushed back from contact with the work by the spring on the stud.

The work is located by the V-blocks in the center, the table is partly cut away to secure a better bed for the piece. The slab cutter operates at the rear of the table, and is quite clear of the operator, who has plenty of room to change the pieces as they come around to the front of the machine.

POINTS FOR CONSIDERATION

The following are a few points to be considered before designing milling fixtures:

1. Will the fixture be similar to some other which has been used to advantage; if so, can you improve upon it?
2. Has the part undergone any previous operations, if so, can you use any of these points to step from?
3. Is it absolutely necessary to work from any previously bored or finished surface?
4. Will a table of clamping schemes help you?
5. Can any stock castings be used for making the fixture?
6. Would it be of advantage to make a special pattern?
7. Can the part be quickly placed in the fixture?
8. Can the part be quickly removed from the fixture?
9. Is the part held firmly so that it cannot work loose, spring, or chatter while the cut is on?
10. Always bear in mind that the cut should be against the solid part of the fixture and not against the clamp.
11. Can more than one part be placed in the fixture, and by so doing increase the output?
12. Can the chips be readily removed from the face of the fixture upon which the part is located or clamped?

13 Are the clamps on the fixture strong enough to prevent them from buckling when they are tightened down on the work?

14. In using cams or wedges for binding or clamping the work, always bear in mind that through the vibration or chatter of the fixture or work they are apt to come loose and cause a great deal of damage.

15. Are there any special wrenches to be designed to go with the fixture?

16. Can you use a reversible key in the milling fixture, and will it fit the millers on which the fixture is to go?

17. Can a gage be designed, or hardened pins added, to help the operator set the milling cutters, or check up the work?

18. Must special milling cutters, arbors or collars be designed to go with the fixture?

19. Is there plenty of clearance for the arbor collars to pass over the work without striking?

20. If the fixture is of the rotary type, have you designed an accurate indexing arrangement?

21. Can the fixture be used on a standard rotary indexing head?

22. Is the fixture strong enough to prevent any vibration while the cutters are in action?

23. Can the fixture be made to take in more than one operation? If so, would it be advisable to have it do so, instead of making two fixtures?

24. Have you, in the designing of the fixture, brought the work as close to the table of the miller as possible?

25. Can the part be milled in a standard miller vise by making up a set of special jaws and thus do away with an expensive fixture?

26. If the part is to be milled at an angle could the fixture be simplified by using a standard adjustable milling angle?

27. Is there any danger of injury to the operator through the faulty design of the fixture?

28. Can lugs be cast on parts to be machined to enable you to hold it?

29. What arrangements have you made to prevent the clamps from turning while they are being tightened on the work?

30. How many different sized wrenches must the operator have in order to tighten all clamps? Why won't one do?

31. Can the work be gaged in the fixture, or must the fixture be cut away so that a micrometer or snap gage can be used?

32. Can you use jack-pins to help support the work while it is being milled?

33. Will a reciprocating fixture be of advantage in getting out production?

34. Can a profile be used to help the operator in accurately locating the part?

35. Have you placed springs under all clamps?

36. In the building up of the fixture, are all parts properly screwed and doweled in place? Is there any danger of their working loose?

37. Are all steel contact points, clamps, etc., hardened?

CHAPTER IX

PNEUMATIC FIXTURES FOR HOLDING WORK *

The illustrations show part of a line of quick-acting fixtures, designed for profiling, milling and drilling metal parts of a well known machine. It had been found that in the fixtures used previously, more of the operator's time was often taken in clamping

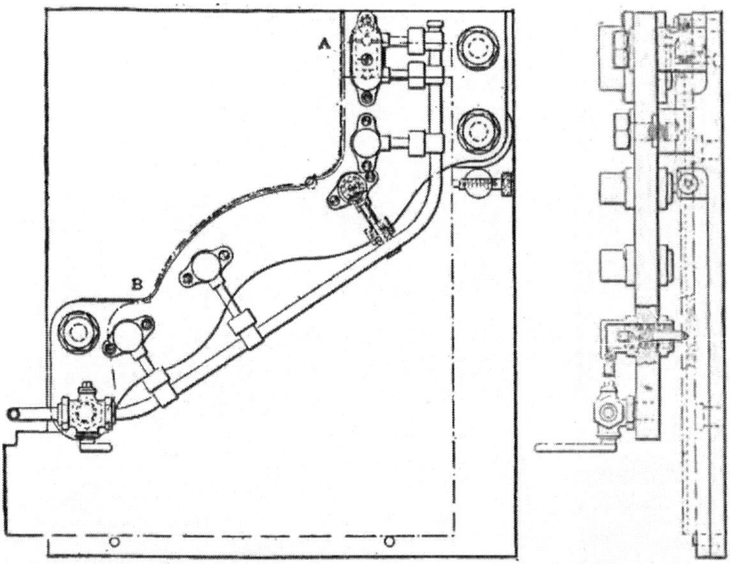

FIG. 223.—PNEUMATIC FIXTURE FOR PROFILE WORK.

and unclamping the work, than was consumed in the actual cutting operations.

There is nothing startling in the design of these fixtures except, perhaps, a little more tool work in the making of them than there would have been if simple screws, clamps, or eccentrics had been used to hold the work in place. But we wish to impress firmly the great saving in time, and consequent cost of manufacture brought about

* M. E. SERVICE. Head of tool designing department, National Cash Register Company.

by the use of fixtures of this nature. The line of fixtures shown herewith is mostly for cutting brass, and as the cuts are made very quickly, the operator will spend the greater part of his time in placing and removing the work. The first fixture, Fig. 223, is a profile fixture for the cut, extending from *A* to *B;* the cut is about 12 ins. long, not very heavy generally, excepting at the top part, a distance of about 5 ins. where sometimes the castings come a little heavy at this place. This fixture consists of a cast-iron base 1 in. thick held to profile-machine platen by two fillister-head screws. On this base, held by three screws, is a machine-steel form plate 13/16 in. thick, which is carbonized and hardened on the tapered form surface at *A B*, on which the taper roll of the cutter arbor shown bears; this is made taper for adjusting the cut as the cutter is ground. The work is slid under this form plate, dropped over and against the two locating pins at the bottom edge, and located

FIG. 224.—ARBOR FOR PROFILE MILLING.

sideways with nurled screw forcing the work against the pin which is cut away to fit the corner in casting as shown, then quickly touching the valve lever the work is clamped.

The fixture previously in use had five 3/8-in. screws for clamping the work. At the rate of 65 cents per hundred for the making of this cut, an experienced man by working very hard, and pushing the job as much as possible, was enabled to make $2.80 to $3 per day of 9 hours. This was replaced by the five 3/8-in. screws with the small plungers as shown, worked by compressed air.

As they were made to clamp close to the working edge, they were made small, 13/16 in. diameter to clear the profile spindle, and making them small, naturally gave very little pressure, but in a profile cut of this nature, the pressure of cutter is nearly all sideways. So the plungers were made with a pointed end like a center punch, and as this point enters the work about 1/32 in. there was no trouble whatever holding the work, the secret of it being the pointed clamps.

With the pneumatic fixture the operator could make the same pay with the price lowered to 30 cents. The air pressure was supposed to be 80 pounds, although at that time the air plant was small and the pressure would sometimes get down to 50 or 60 pounds. Fig. 224 shows an arbor for cutters used to follow a form in profiling.

HOLDING A WEAK PIECE OF WORK

Fig. 225 is a profile fixture using a form cutter for the cut on surface marked *x x* in section, and extending around from *o* to *o* on a brass-plate casting of a shape as shown by the dot-and-dash line. This casting is narrow and weak at and near the ends *o o*. The fixture has a cast-iron base 1¼ ins. thick, which is screwed to the profile platen. A machine-steel form plate with hardened straight-faced contour, used for guiding the cutter, swings on hinges to admit of placing the work, these hinges being at the right-hand side of the operator. This form plate is further guided and controlled by having two slots ½ in. wide, which pass over, fitting sideways and bearing against the shoulder of the hardened-steel posts, as shown at *A A*. Inserted into clearance holes in this form plate are several bronze castings, being held to form plate with fillister screws.

These castings have a number of 11/16-in. holes in them for operating the air pistons, which work through levers, as shown in section view. This also shows the center punch points and how they get close to the outer edge of the work. In addition to the cylinders operating the clamping points, there are two other ⅝-in. cylinder castings which operate the lock bolts holding the form plate to base of fixture, as shown at *A A*. The clamping cylinders are all connected by holes and piping to a small piece of rubber tube which leads to the valve in the corner of the base of fixture. The two form-plate locking cylinders are connected to another piece of rubber tube, which also leads to the valve. The form plate, being heavy, is raised and lowered for the admission of work by a very simple air hoist, which is not shown. The work which is shown finished, in a dot-and-dash line, has a hole at *A A* which fits over locking post, and is swung against the pin at *o* for locating, the form plate being up at this time. Touching the air-hoist valve the form plate quickly and lightly drops to position. Next, touching the valve on corner of fixture first admits air to locking cylinders, and second, with the same movement, admits air to all the clamping cylinders holding your work. To remove work, touch valve on corner of fixture, then touch hoist valve and take work out.

In the fixture in use previously the work was held by clamps and screws, without any lid, the work being slid under the form plate

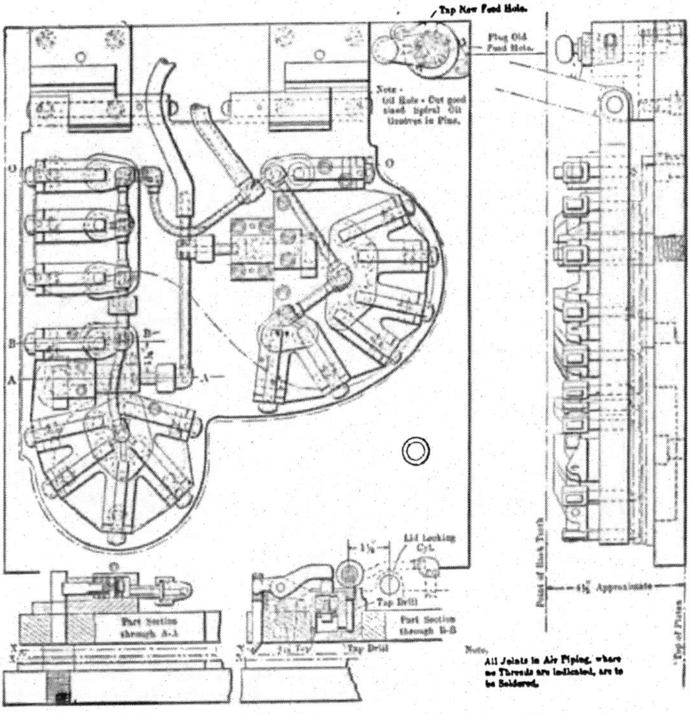

FIG. 225.—PROFILE FIXTURE FOR FORM CUTTER WORK.

and clamped, and with this style of fixture $1.80 per 100 was paid for the making of this profile cut, but with the fixture as shown, using air, the price paid is 90 cents per 100, and this enabled the same man to make the same wage as before.

ANOTHER PROFILE FIXTURE

Fig. 226 shows another style of profile fixture, for making some boss cuts at the back of fixture and a surface cut at the front of fixture as pointed out on the drawing. A right and left brass side plate, is shown by the dot-and-dash line in this fixture, although in use only one side plate is put in the fixture and operated on at a time.

The base of this fixture is cast iron $2\frac{3}{8}$ ins. thick, cored out to lighten and with $4\frac{5}{8}$ x $1\frac{1}{2}$ ins. diameter, counterbored holes for the

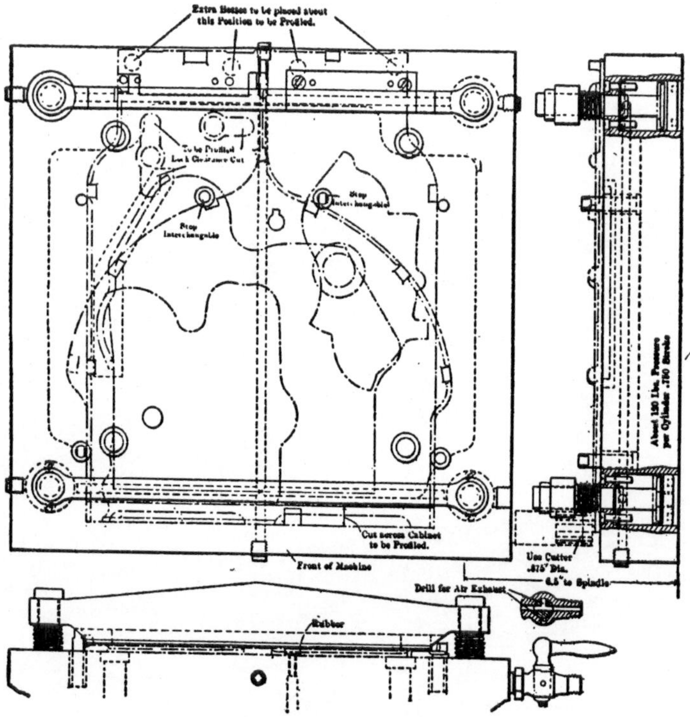

FIG. 226.—PROFILE FIXTURE FOR DIFFERENT CUTS.

pistons to work in. The pistons in this case work at the bottom of the open cylinder hole, the cupped-leather packing being held in place by a washer and round nut underneath forcing against the shoulder of the round steel-headed screw which connects piston to clamping plate. This screw, or piston-rod is automatically packed at the top of cylinder bore by a reverse cupped-leather packing, held in place by a metal washer and screws as shown in section. When the air is shut off, the pistons are brought to their normal position by the two springs under each clamping plate.

The four cylinders are all connected together by three 1/4-in. holes drilled through the cast-iron base from side to side with a long shank drill, and having the outer openings closed with plug as shown. The steel clamping plates are slotted at the bottom with a 1/4-in. width of slot in which is forced a long strip of medium hard rubber 1/4 x 1/2-in. for clamping. This rubber is used to enable the clamp to accommodate itself to irregularities in thickness of work, and is also used largely to prevent marring any surface which is stippled, or where a clamp mark would show in the final finish. The valve in this case is shown in front at the right-hand side of the fixture.

FIXTURE FOR DOUBLE-HEAD MILLING

Fig. 227 shows a double-head mill fixture for milling simultaneously both sides of a brass casting about 1/2 in. square of a section as shown, and of five different lengths of about 9, 16, 18, 23 and 26 ins. The drawing shows the work, with cut finished, but in position ready to be clamped by a touch of the valve lever at the top of fixture. This fixture is about 34 1/2 ins. long over all, has a cast-iron base for clamping to platen, with a hardened machine-steel strip for supporting the work. The work is put in from one side against the adjustable stop arms, which are thrown into a fixed stopped position by the torsion spring surrounding the stop-arm carrying shaft. These arms have stop pins and adjustable screw stops for different widths of work to be milled, and are always in the stop position when work is ready to be put in.

Fastened with cap-screws to the base casting, is a cast-iron yoke which is drilled and reamed for eight 1-in. pistons, seven of these pistons having tool-steel hardened center-punch points for clamping the work. The eighth cylinder is used to push the stopping arms out of the way of the cutters. The cutters are shown in the starting position, and have a tendency to hold the work down and also against the stop pin shown.

The air enters the valve through a short piece of rubber tubing, and when the valve is opened, the air passes to the seven clamping cylinders through a single 1/8-in. brass tube. The air also exhausts through this same tube, to a by-pass in the valve in releasing. Directly above each cylinder, this air-supply tube is punctured with a small hole packed with a piece of soft rubber and the air tube clamped at this point with the screw and plates shown. Four of these cylinders have air cut-off screws at the side to save air when doing the shorter work. A second small brass tube leads from the valve to the end cylinder operating the adjustable stop arms previously mentioned. In operation, the stop arms being in position, the piece to be milled is placed in the fixture and the valve lever is moved.

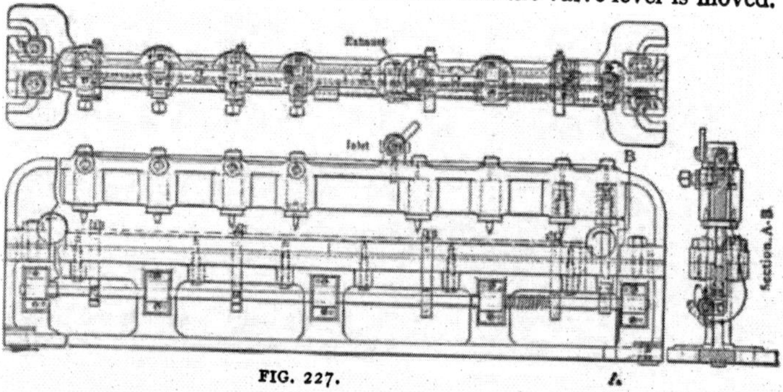

FIG. 227.

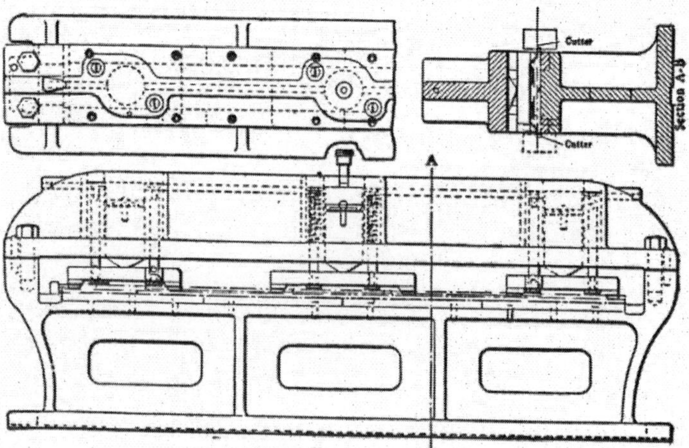

FIG. 228.

FIGS. 227 TO 228.—FIXTURES FOR DOUBLE-HEAD AND STRADDLE MILLING.

This single movement admits air first to all the clamping cylinders you wish to operate, and second to the cylinder removing the stopping arms from the path of the cutter. To release the air you move valve lever back. The use of air in place of screws cut the price of this straddle cut from $1.25 to 60 cents per hundred, with the same wages earned in both cases.

WHERE STRADDLE MILLS ARE USED

In Fig. 228 is shown a double-head milling fixture for making at one operation the cut shown in section at *x* and *x*. This cut is on a brass casting, and is nearly 26 ins. in length, the thickness of cut being not over 5/32 in. The work in these fixtures stands high, about 6 ins. from the platen of the machine; the reason for this is to keep the adjustment of the cutter heads in practically the same up and down position for the operator, as in some of these fixtures for different cuts, the cutters must cut from underneath. The base of this fixture is cast iron with hardened bearing strips at the side for clamping the work on, the work being located sideways against stop pins on the border of work beneath.

The cast-iron yoke is fastened to the base by screws and pins, is bored out in three places on its underneath side for 1¾ in. pistons, which are simply cold-rolled pieces of shaft tapered on the lower end to bear against clamping plate, and having the usual cupped-leather packing held by a washer and screw at the top. At the side of each of these cylinder holes are two ⅜-in. holes drilled and counterbored from the top to ¾-in. diameter, for a coiled spring, connection of spring and clamp being made by the rod with screw in each end. The valve in this fixture is placed at the connection shown in the center, and has a rubber tube connecting to air-supply pipe. The air connection to the three cylinders is through a ¼-in. hole drilled from each end of yoke, with the ends of holes plugged.

HOLDING TWO PIECES AT ONCE

Fig. 229 shows a fixture for making a hinge cut on each end of a curved brass cabinet lid of different lengths, the longest lid being about 19 ins. The fixture is designed to do two lids of the same length at a time, both hinge cuts being made simultaneously with a double-head mill, and the cutters being double inter-locking with herring-bone teeth, this being done to make a smooth cut.

This fixture being made to do different lengths of lids, it is made with two clamping heads, made to slide toward the center of fixture, guided by a tongue, and held down to the cast-iron base plate by two bolts sliding in the T-slot in the base. On these clamping heads, a machine-steel forging, planed to fit the contour of the lids to be milled is fastened by screws. These lids come against case-hardened heads, formed on this forging at the top. This forging is stiffened by a plate, which is tongued and screwed to it, and further fastened to the clamping head at each side.

Each clamping head in a recess cored in each side has a cast-bronze cylinder of 2⅛-in. bore, fitted with the usual leather-cupped packed piston, carrying a clamping plunger 1¾ ins. wide at the bottom which is beveled, forming a wedge between the work to be milled, and two hardened pins opposite. The work, which is shown clamped in the fixture in a dot-and-dash line, is previously milled on the lower and upper edges.

To release this work, simply touch the valve lever, shown on the base in front at the center which raises all four wedges, this being all there is to the clamping or unclamping of these two cabinet sides. The air-connecting holes are all made by drilling through the base and heads, leading to the four cylinders, and to the uninitiated, it is something almost uncanny to see a fixture of this kind clamp work without any visible connecting link. Attention is called to this peculiar method of clamping for such a heavy cut as this, as, from the end, it seems as though a cutter would push the other side away; but, taking into account the point at the clamping wedge it would have to swing from, it will be seen that it only holds the tighter.

HOLDING TIE BARS OF DIFFERENT LENGTHS

Fig. 230 is a fixture designed to mill the ends of two cast-iron tie bars at one time, and also to mill bars of different lengths. The drawing shows, by a dot-and-dash line, two of the longest bars clamped in position. This fixture has a cast-iron base carrying four movable cylinders in which are the usual pistons as seen in the section, each piston having a steel plunger pointed and hardened on the end, which pushes against a lever that is keyed to a forged tool-steel shaft ¾ in. diameter, swinging in bearings, and having on the other end another lever forged on it. This lever has a knife edge hardened and bears against the work to be milled.

The work to be milled is supported at each end by a machine steel, carbonized and hardened block which is adjustable for the different lengths of tie bars to be milled, and is located with a tongue, and held to base by a single bolt. This bearing block is easily removed for other shapes suiting different work. The valve in this fixture is on the base and is connected to the four cylinders

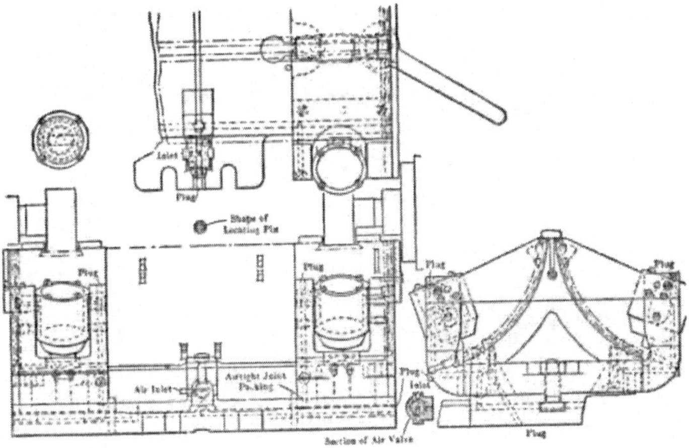

FIG. 229.

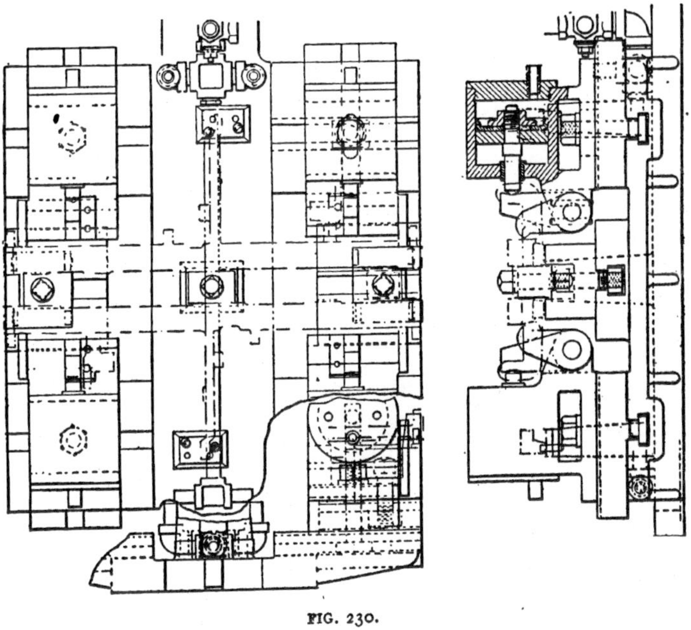

FIG. 230.

FIGS. 229 AND 230.—TWO OTHER PNEUMATIC FIXTURES.

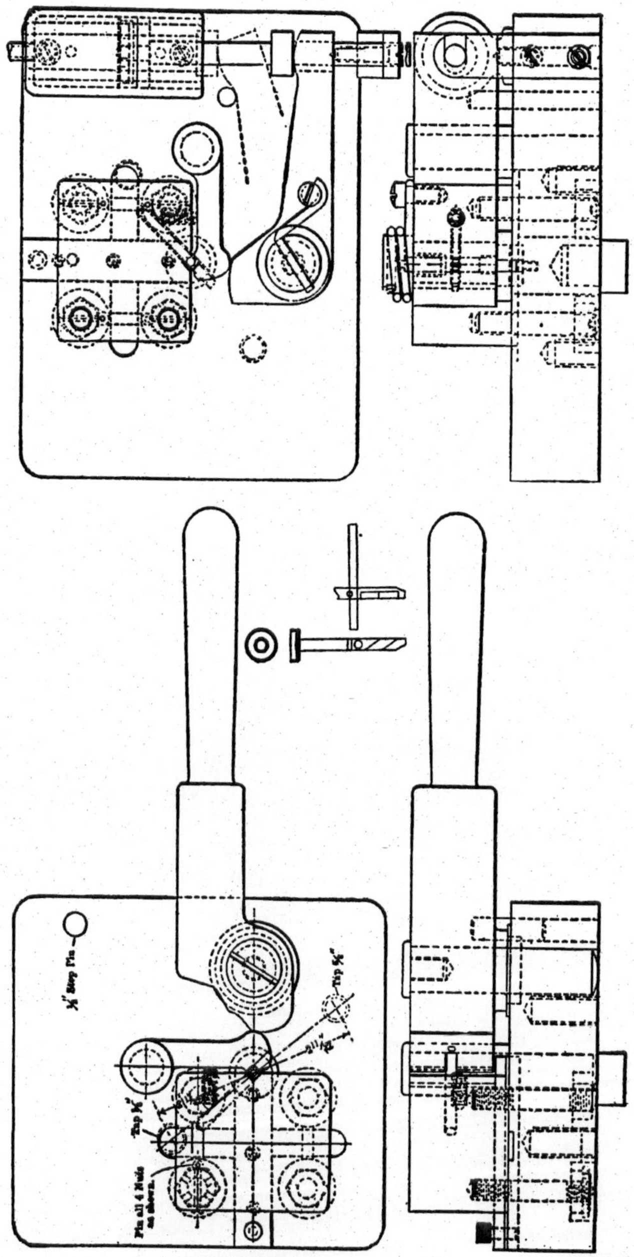

FIGS. 231. FIG. 232.

FIGS. 231 TO 232.—HAND AND PNEUMATIC FIXTURES FOR THE SAME PIECE.

by short pieces of rubber tube which are joined to connections leading to a hole in the base.

These cylinders are made large, 3 ins. diameter, as the cut is heavy, and the force of it comes directly against one of these bars, which must be counter-acted by the air pressure. A single touch of a lever clamps these two tie bars, and a touch of the same lever releases them.

BEFORE AND AFTER TAKING AIR

Figs. 231 and 232 show an example of before and after taking the air. In Fig. 231 is a fixture for riveting the round key cup to the round-shouldered end of a square-steel key. This key has a slot milled near the bottom end, and also a hole with a pin projecting from it near the center, making it weak at this lower end, as will be seen by looking at the detail of this key shown at the side of Fig. 232. These keys being weak must have a powerful clamping pressure applied to them to prevent them from springing by the action of the riveting machine forcing them down on their lower end. These keys are held in a V with a V-shaped lever being clamped with a cam lever about 10 ins. long. It is a simple movement to operate this lever, but it is tiresome and slow in comparison with the mere touch of an air valve. Fig. 232 shows how this fixture was doctored, by sticking a little air cylinder in one corner, connecting its piston-rod to a cam-shaped lever, when a light touch of the valve lever clamps or unclamps the work. This increased the output, and also insured the same clamping effect, preventing any springing of the keys brought about by the operator not putting sufficient clamping effort on the cam lever, shown in Fig. 231.

AIR APPLIED TO LARGE FIXTURES

Fig. 233 shows how air was applied to several large jigs for drilling brass cabinet sides. The outline of one of these jigs is shown in a faint line, but it does not show any of the 13 bushings in the top or lid of this jig, or the four bushings for the base holes in cabinet side, located in the back side of the jig. There was also one bushing hole in the side opposite the lid hinges. These 18 holes are all drilled simultaneously on an automatic multiple-spindle drill press. This jig originally had five nurled headed screws in the lid for clamping the work, and also had screws in the lid-locating posts, which are fastened to the base; these screws clamped and unclamped the lid to the base of the jig, by a quarter turn. When using these screws \$1.25 was paid for the drilling of 100 cabinet sides, but after applying three air-lid locks, and the nine air-clamping

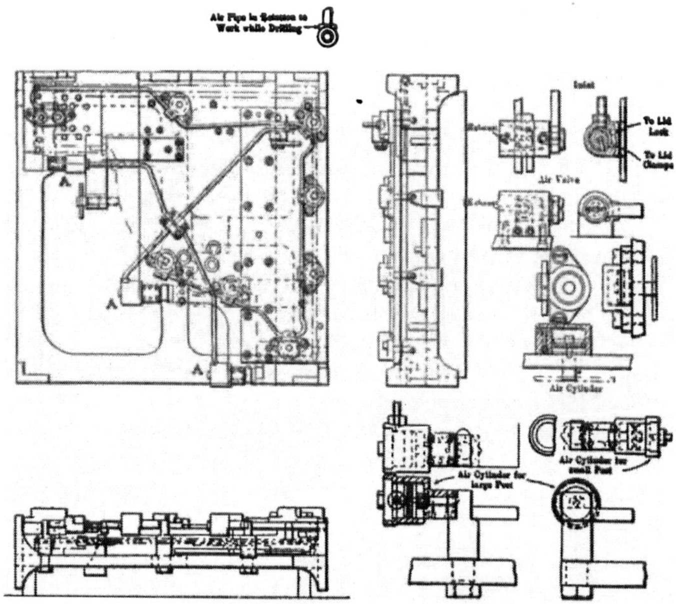

FIG. 233.—AIR APPLIED TO LARGE FIXTURES.

cylinders as shown, which all work with only a touch of a valve lever, the price was cut to 52 cents per 100, with the same man on the same machine earning the same wages in each case.

In these drill jigs a standard cast-bronze cylinder casting was used, being faced off on the end which is screwed to the top of the jig lid, this faced end having a projection in which is milled a ¼-in. slot round at the bottom and in this slot is the 3/16-in. outside diameter, brass air-supply tube, which is twined around like a string of putty to any cylinder you have in place and is then punctured with a 1/32-in. hole, packed around with soft rubber, the connection being held air tight with the two fillister-head screws which also hold the cylinder. The details of the clamping cylinders, lid-locking cylinders and the air valve are all clearly shown in the sectional detail. These clamping cylinders, lid locks, valve and piping are all on the top of the lid jig, the valve is in one corner, and is connected to the air supply pipe by a rubber tube.

To operate this jig, place the cabinet side in position, lower the lid, and touch the valve lever. This allows air to pass through the line of pipe to the three lid-locking cylinders *A A A*, and with only the one movement allows air to pass through the other line of piping

FIG. 234.—PNEUMATIC MILLING FIXTURE.

connecting the nine clamping cylinders. The jig is then ready to roll to position under the 18 revolving drills. A backward movement of the valve lever unclamps the work, and then unlocks the lid.

These fixtures are rapid, hold the work well, and in any of the departments where we place fixtures of any nature operated by air the workmen like them, and the foremen naturally become so enthusiastic about them that in some cases they clamor to have all fixtures operated by air.

A PNEUMATIC MILLING FIXTURE

One of the milling jigs used by the Cincinnati Milling Machine Company is shown in Figs. 234 and 235. The first shows the general arrangement of the cylinders and jaws while the latter gives details of construction. There are six cylinders, three on each line of air pipe and in addition to cutting out either set of three, each cylinder has a separate cut-out in the shape of a plug cock. The pistons each have four packing rings.

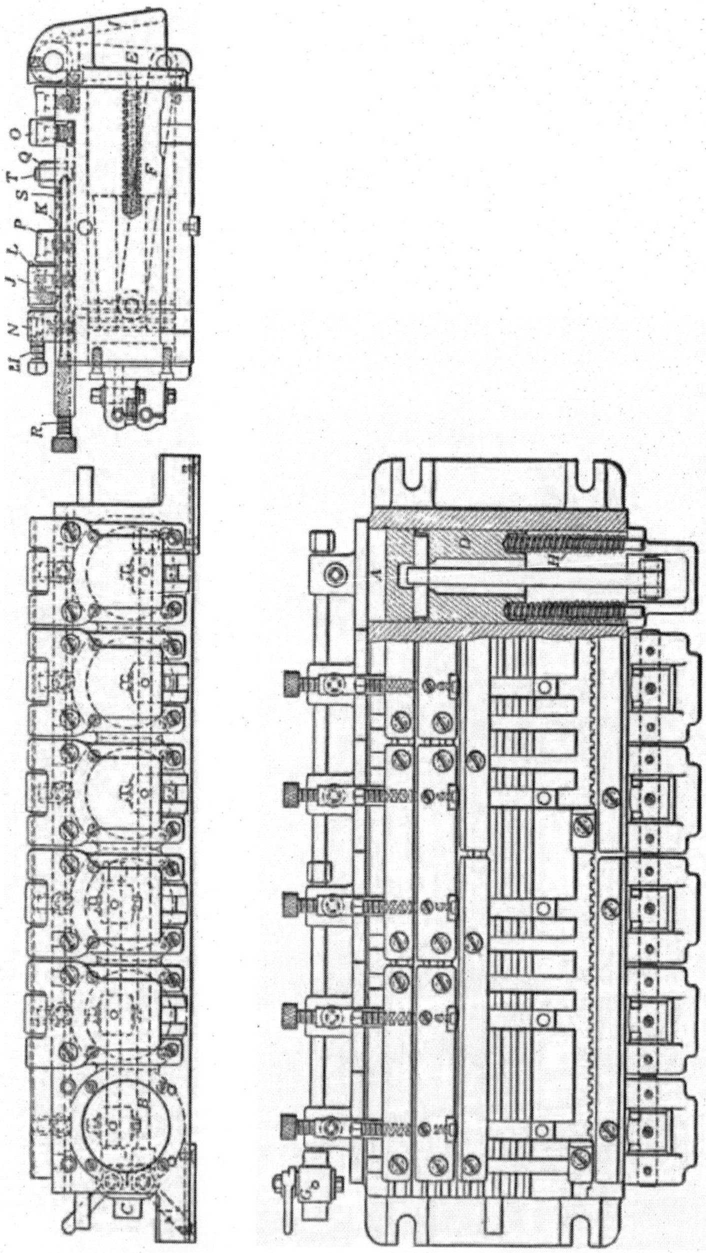

FIG. 235.—DETAILS OF THE SIX CYLINDER PNEUMATIC MILLING FIXTURE.

The operation of the vise is as follows: Air at 80 pounds pressure per square inch is admitted to the chamber *A*, through the pipe *B* and cock *C*, and forces in the piston plunger *D*. Motion is transmitted to the movable jaw *E* through the link *F*, causing the jaw to clamp the work securely.

To release the work the air cock is closed, and the pressure against the piston is relieved by the air escaping through the exhaust opening *G*, and the piston and jaw are returned to normal position by the springs *H*. *I* is a cast piece serving as a cover for the working parts of the jaw.

It will be noticed from the construction of the air cock that either three or six of the jaws can be operated, depending on the size of the work. *J* is the back or fixed jaw. It is fastened in the slots *K*, and can be adjusted back and forth. Hardened pieces *L* are provided for bearing against the work. The set screws *M* are carried in the backing strips *N* and are tightened against the fixed jaw to give it greater rigidity. This work is positioned lengthwise against the adjustable stops *O*, and is supported by the bearing strips *P* and the blocks *Q*, which latter are provided with pins *T*, the height of which is adjustable by the screw *R* and pin *S*.

This vise, with its various adjustments, covers a wide range of adaptability, and many pieces of varying size can be handled in it.

One point to be borne in mind regarding pneumatic fixtures is that the work must be timed so as to insure the completion of a cut once started before the machinery is shut down for the day, or else the fixture will release the work in the machine, and the piece will have to be done over, as it is practically impossible to replace the work in exactly the same position in the fixture.

CHAPTER X

USES AND ADVANTAGES OF THE LATCH JIG *

Of all the special tools used in a modern manufacturing establishment for the economical production of a large number of small parts, no tool offers in its scope a wider adaptability for the rapid production of interchangeable parts than the latch jig. In fact, no tool offers a more satisfactory subject to the tool designer than this particular type of tool.

The evolution and refinement of the latch jig has been accomplished through urgent demands for a jig that must be simple in construction; capable of producing uniformly accurate work; designed to withstand rough usage; rapid and easy of manipulation; easily cleaned, and accessible when clamping and removing the work.

All of these essentials are embodied in this particular type of tool. One movement is sufficient to open it and expose its gage points and nest. It is easily and quickly cleaned. It is conveniently and quickly loaded. It is simple in its construction, hence capable of withstanding rough usage and still maintain its accuracy. One movement and the cover is closed and the work clamped securely in position. Speed, the most desirable feature of a good tool, makes the quick-acting latch jig a most desirable factor in good tool design.

Referring to Fig. 236, the component parts of this tool may be enumerated as follows: *A*, Frame; *B*, cover; *C*, cover hinge pin; *D*, latch; *E*, latch hinge pin; *F*, latch fulcrum pin; *G*, latch stop screw; *H*, latch stop-screw nut.

It is not feasible to use a latch jig on work that varies greatly in thickness, such as rough castings and rough forgings, without providing a compensating device on the cover. This is due to the fact that the cam on the latch is short, and provides approximately 0.03 for variation in the thickness of the work.

Generally speaking, there are three styles of this type of jig, and they may be designated as follows: The open type, the box type and the reverse type. The open type, with handle and latch pro-

*By J. N. Bethel, Superintendent, Taft-Pierce Manufacturing Co., Woonsocket, R. I.

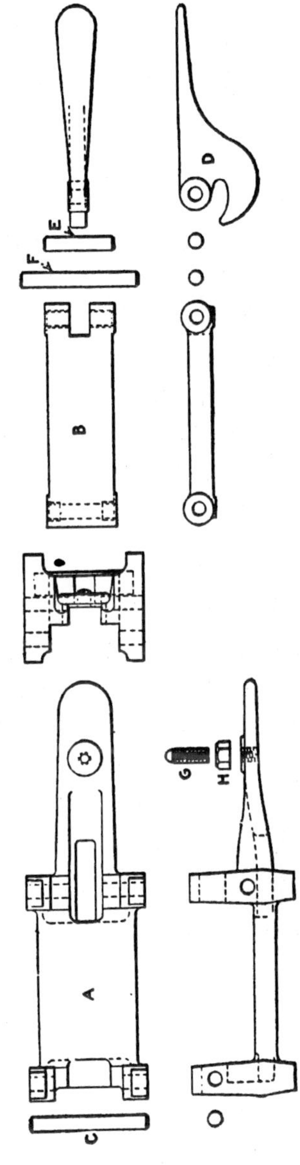

FIG. 236.—PARTS USED IN A LATCH JIG.

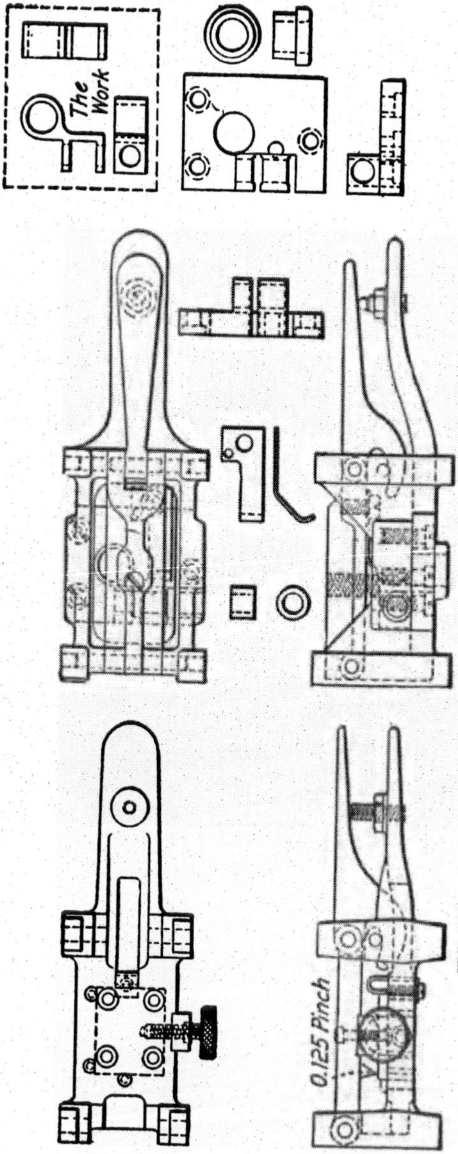

FIG. 238.

FIG. 237.

FIGS. 237 TO 238.—LATCH JIGS FOR FLAT WORK.

jecting, as illustrated in Fig. 237, is particularly recommended for flat work, which requires plenty of finger room. The gage points and nest are conveniently located. This type is recommended where holes are to be drilled and reamed at right angles with the side of the work only.

When the parts to be drilled are of uniform thickness, if so desired, the bushings may be located in the cover. It is good practice to

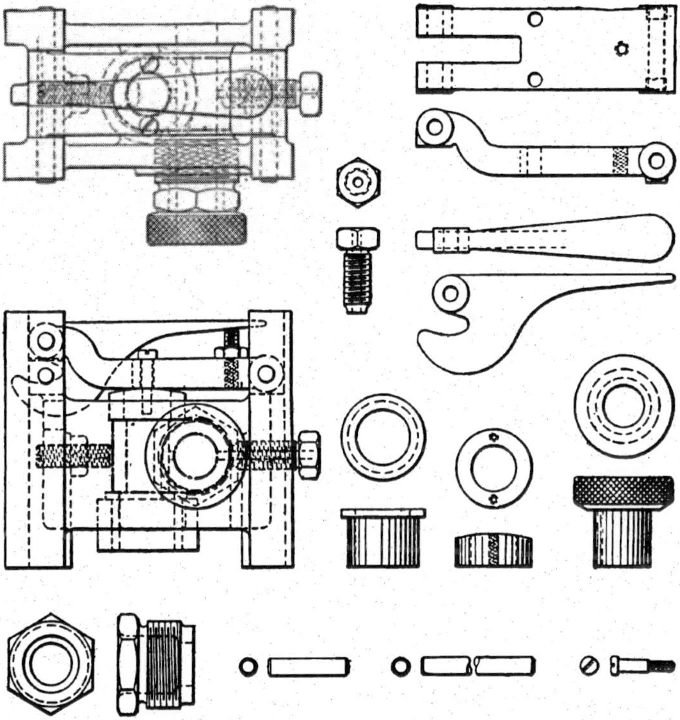

FIG. 239.—CONSTRUCTION OF LATCH JIGS.

do this whenever possible, as it dispenses with the necessity of turning the jig over when preparing for the actual operation of drilling.

If the work varies considerably in thickness, the cover may be clamped on a finished seat at *A*, Fig. 236, and a binding screw added to the cover for clamping the work. By this method the center line of the bushing will be presented at right angles with the nest or gage seat which receives the work.

In addition to the open type, an illustration is given in Fig. 238 of the inclosed or box type, which permits the use of bushings in the sides as well as in the cover and bottom. This jig is capable of

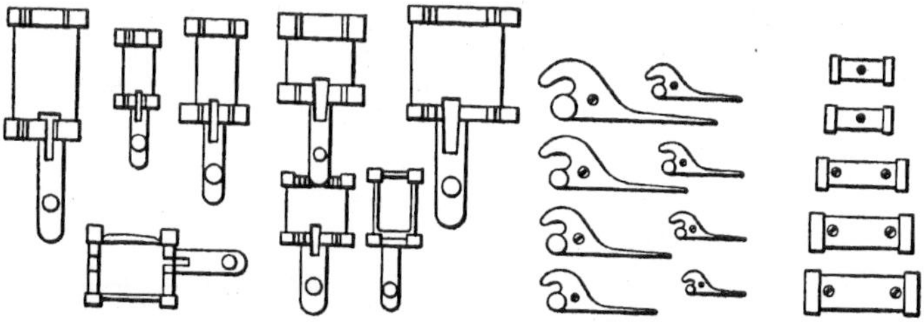

FIG. 240. FIG. 241.
FIGS. 240 TO 241.—STOCK PARTS FOR LATCH JIGS.

accommodating a diversified line of parts with very satisfactory results.

Frequently it is necessary to drill a part from five sides. To do this, a very ingenious arrangement is used, as in Fig. 239, whereby

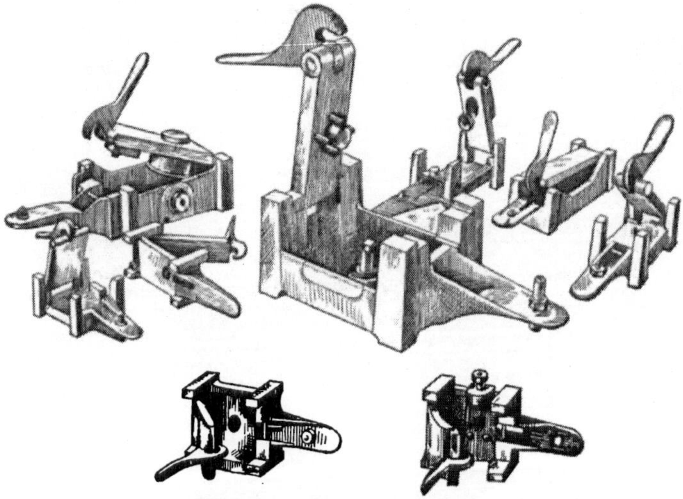

FIG. 242.—A GROUP OF LATCH JIGS.

the handle is dispensed with and the latch is reversed to swing toward the center of the jig, using an offset cover to accomplish this purpose. This jig is capable of a very wide range of work and gives excellent results.

In the Taft-Pierce Manufacturing Co.'s very elaborate and complete stock of standard jig, fixture, punch and die and miscellaneous tool fittings, there are kept eight sizes of jig frames and fittings in steel castings. These are being constantly used and adapted to the needs of the rapid and accurate production of parts for adding machines, typewriters, sewing machines, cash registers, parts for small automatic machinery, motorcycle motors, automobile parts, etc. These stock-jig parts are represented in Figs. 240 and 241, and a group of completed jigs is seen in Fig. 242.

Where large quantities of work are involved, it is good practice to make the frame of a steel casting and harden the feet. This is best accomplished by machining the body of the jig before boring to receive the drill bushings. Drill the holes to receive No. 4 cover hinge pin and No. 6 latch fulcrum pin, then pack the frame in an iron box and cover the work with any of the carbonizing preparations. Seal the box to make it air tight and heat the work in a hardening furnace to about 1500 to 1600 deg. F. for a period of six to eight hours, depending upon the size of the frame. Remove the box from the furnace, keep it sealed to exclude the air, and allow the work to cool in the box.

Next, heat the ends of the feet for a distance of about ¾ in. in the lead pot to a cherry red and quench in cold water. Surface grind the legs, the feet and the seats square with each other, then bore for the drill bushings, registering from the hardened and ground surfaces. By pursuing this method, an accurate and extremely durable jig is obtained.

CHAPTER XI

DESIGNS OF AND MATERIALS FOR GAGES

A thorough system of gaging is the main basis of interchangeable manufacture, and no firm can enter successfully into that field without giving gages in general and limit gages in particular a liberal consideration.

It is supposed in many quarters that gages are such expensive articles that their use should be reduced to the lowest possible limit. This is an erroneous supposition and anyone who has thoroughly investigated the subject, or who is responsible for the economical production of a number of interchangeable parts, knows that a liberal sum spent on gages is the cheapest way in the long run.

The grinding machine, with its extreme accuracy in producing true cylindrical and parallel surfaces, has eliminated a great deal of the cost that at one time made gages so expensive.

For manufacturing purposes lapping is not necessary; a gage direct from the grinder should be sufficiently accurate for all practical uses. Efforts have been made to economize in gages by the use of micrometers, because of their range of uses, but the micrometer or vernier is not a manufacturing gage at all, either as regards accuracy or rapidity. No two men read it alike, even in a toolroom, and the difference is much greater in the manufacturing plant.

The time taken to read and adjust the micrometer is slow and an operator generally likes to apply it several times before he is sure of his reading; even then he possibly has his doubts as to whether the inspector will agree with him when the piece comes to be tested later on.

All manufacturing plants, practicing to any degree interchangeable manufacturing, should have a set of reference gages, with which all the gages used in the shop should be regularly compared. It would pay to have a special department to see that all gages were kept accurate and to condemn and make renewals when necessary.

Any extra expense entailed would be much more than counterbalanced by the saving effected at the assembly benches, when the uniformity of the parts would allow of their being put together with greater rapidity.

SNAP GAGES

This type of gage occupies a very prominent place in works of manufacturers who are desirous of maintaining a standard of interchangeability in their product. To obtain the best results a set of reference plug gages should be kept, to which the snap gages should be referred from time to time, and any wear immediately corrected. This can be done by closing in the gage to an amount slightly in excess of the wear, and regrinding or lapping to exact size.

This closing-in process in no way injures the gage, if carefully done, and makes it possible to use the same gage almost indefinitely. Snap gages are much more sensitive than is generally supposed; in using one no pressure should ever be applied to force it on the work. It is not necessary.

With a pair of snap gages having a difference of 0.00025 in. the smaller size will refuse to pass over the work when the larger size has done so with an appreciable amount of shake. This is probably closer working than an ordinary micrometer can be relied upon to give.

Snap gages may be designed in a great many ways to suit special conditions. The illustrations show a number of these of a more or less standard form. Figs. 243, 244 and 245 are self-explanatory, illustrating only three different methods of making them. The first is made of plate stock, the second of round or bar stock, and the third of square bar stock. The working surfaces should be hardened and ground.

A method of adapting these gages for use on the lathe with the mandrel in place between the centers is shown in Fig. 246. Attention is called to the bend or offset in the center of the gage.

A flat steel gage with plenty of clearance is shown at Fig. 247. This, like Figs. 243, 244 and 245, is made of flat plate stock, which is frequently used in the construction of small gages. A hole is shown in the upper right hand corner which allows the gage to be hung up while not in use.

The well known caliper gages are shown in Figs. 248 and 249. They are steel stampings, case hardened and ground to size on the testing surfaces. The renewing of the wearing surfaces of snap gages is often desirable, and Figs. 250 to 254 show various methods

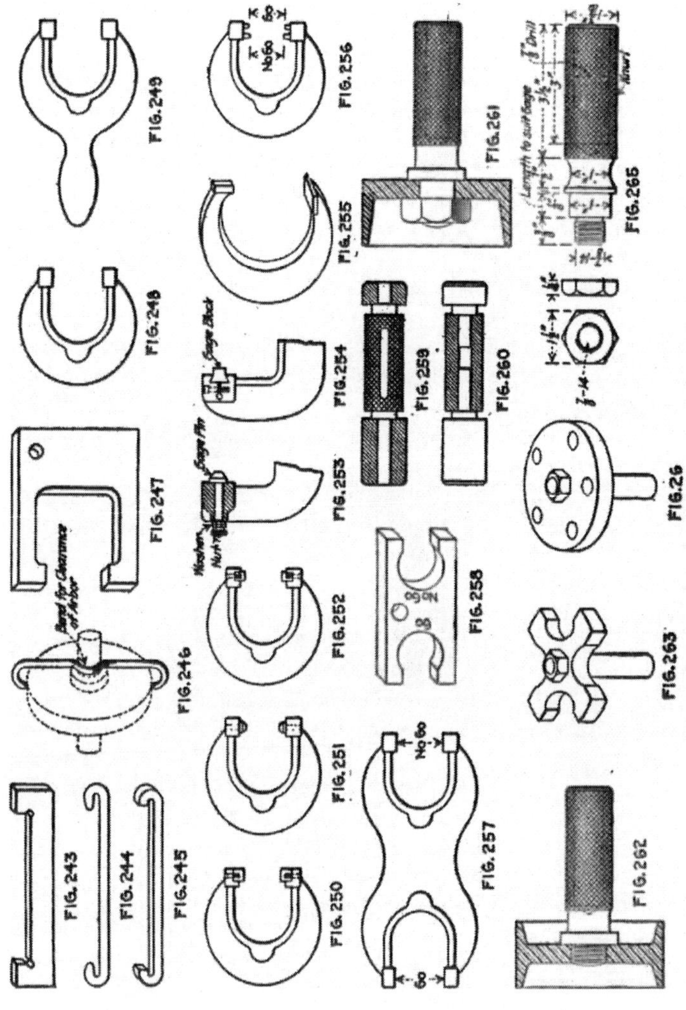

FIGS. 243 TO 265.—GAGES FOR A VARIETY OF WORK.

by which this part may be made detachable from the main body of the gage. The test pieces can be renewed and replaced at slight cost. They are made of tool steel, hardened and set into the forging, after which they can be ground to the required size. Two good methods of making snap gages with interchangeable gage pins or

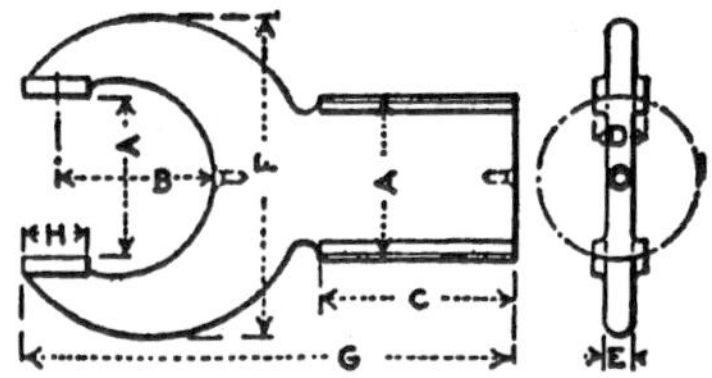

A	B	C	D	E	F	G	H
1/2	7/16	5/8	1/4	3/32	1 1/8	1 5/8	1/4
5/8	9/16	7/8	1/4	3/32	1 3/8	2 1/8	5/16
3/4	5/8	1	9/32	3/32	1 1/2	2 5/16	5/16
7/8	3/4	1 1/4	5/16	3/32	1 7/8	3	3/8
1	7/8	1 5/16	5/16	3/32	1 15/16	3 1/8	3/8
1 1/8	7/8	1 3/8	3/8	3/32	2 1/8	3 1/4	3/8
1 1/4	1	1 1/2	3/8	3/32	2 3/8	3 5/8	7/16
1 3/8	1 1/16	1 5/8	3/8	3/32	2 5/8	3 7/8	7/16
1 1/2	1 1/8	1 3/4	3/8	1/8	2 3/4	4 1/8	7/16
1 5/8	1 3/16	1 7/8	3/8	1/8	3	4 3/8	7/16
1 3/4	1 5/16	2	7/16	1/8	3 1/4	4 5/8	1/2
1 7/8	1 3/8	2 1/16	7/16	1/8	3 3/8	4 7/8	1/2
2	1 1/2	2 1/8	7/16	1/8	3 5/8	5 1/8	1/2
2 1/8	1 9/16	2 1/4	7/16	1/8	3 3/4	5 3/8	9/16
2 1/4	1 5/8	2 3/8	1/2	1/8	4	5 5/8	9/16
2 3/8	1 3/4	2 7/16	1/2	1/8	4 1/4	5 7/8	9/16
2 1/2	1 7/8	2 1/2	1/2	1/8	4 1/2	6	9/16
2 3/4	2	2 5/8	1/2	5/32	4 3/4	6 1/2	5/8
3	2 1/8	2 3/4	1/2	5/32	5	6 3/4	5/8

All dimensions in inches.

TABLE 20.—DOUBLE-END SNAP GAGE SIZES.

gage blocks are shown in Figs. 253 and 254. By making these pins or blocks of various lengths, the cost of a given number of gages can be greatly reduced.

A style of snap gage not very often seen in use is shown in Fig. 255. It is used for testing over the tops of a threaded shaft. The wide surface on one arm being necessary to insure that the gage is being held parallel with the axis of the work. Table 20 gives the proportions of a set of standard combination external and internal gages. The sizes given are for gages made of steel forgings. Sizes larger than three inches should be made in two parts for convenience in handling.

Table 21 gives the proportions of a set of standard snap or caliper gages. The sizes given are for gages made of steel forgings. In some instances the testing surfaces are made of tool steel and

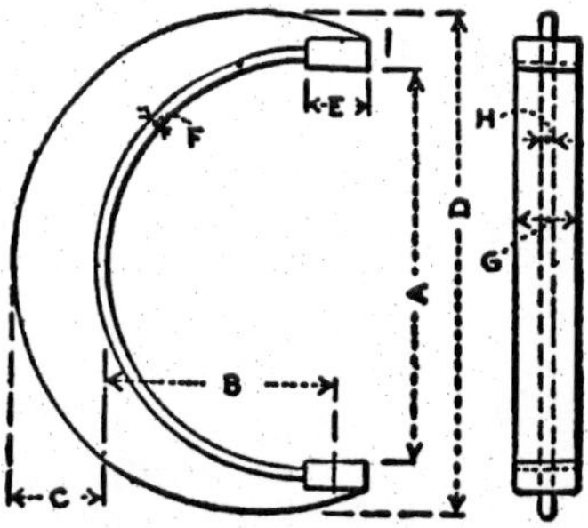

A	B	C	D	E	F	G	H
3	$1\frac{13}{16}$	$\frac{3}{4}$	4	$\frac{1}{2}$	$\frac{1}{8}$	$\frac{1}{2}$	$\frac{5}{32}$
4	$2\frac{5}{16}$	$\frac{7}{8}$	$5\frac{1}{8}$	$\frac{9}{16}$	$\frac{1}{8}$	$\frac{1}{2}$	$\frac{5}{32}$
5	3	1	$6\frac{1}{4}$	$\frac{5}{8}$	$\frac{1}{8}$	$\frac{5}{8}$	$\frac{5}{32}$
6	$3\frac{7}{16}$	$1\frac{1}{8}$	$7\frac{1}{4}$	$\frac{11}{16}$	$\frac{1}{8}$	$\frac{5}{8}$	$\frac{3}{16}$
7	4	$1\frac{1}{4}$	$8\frac{1}{2}$	$\frac{3}{4}$	$\frac{5}{16}$	$\frac{3}{4}$	$\frac{3}{16}$
8	$4\frac{1}{2}$	$1\frac{3}{8}$	$9\frac{3}{4}$	$\frac{13}{16}$	$\frac{3}{16}$	$\frac{3}{4}$	$\frac{3}{16}$
9	5	$1\frac{1}{2}$	$10\frac{1}{2}$	$\frac{7}{8}$	$\frac{3}{16}$	$\frac{3}{4}$	$\frac{1}{4}$
10	$5\frac{1}{2}$	$1\frac{5}{8}$	12	$\frac{15}{16}$	$\frac{1}{4}$	$\frac{7}{8}$	$\frac{1}{4}$
11	6	$1\frac{3}{4}$	13	1	$\frac{1}{4}$	$\frac{7}{8}$	$\frac{1}{4}$
12	$6\frac{1}{2}$	$1\frac{7}{8}$	$14\frac{1}{4}$	$1\frac{1}{8}$	$\frac{5}{16}$	1	$\frac{5}{16}$

All dimensions in inches.
TABLE 21.—SNAP GAGES.

electrically welded to a machine-steel body, after which they are hardened, ground and lapped to the required size.

LIMIT SNAP GAGES

A limit snap gage gives a man something definite and can be rapidly applied. When used in conjunction with stops in the machine it should not be necessary to apply it more than once.

Good examples of what are called limit snap gages are shown in Figs. 256, 257 and 258. This style of gage is used more frequently in the inspection department than in the manufacturing department.

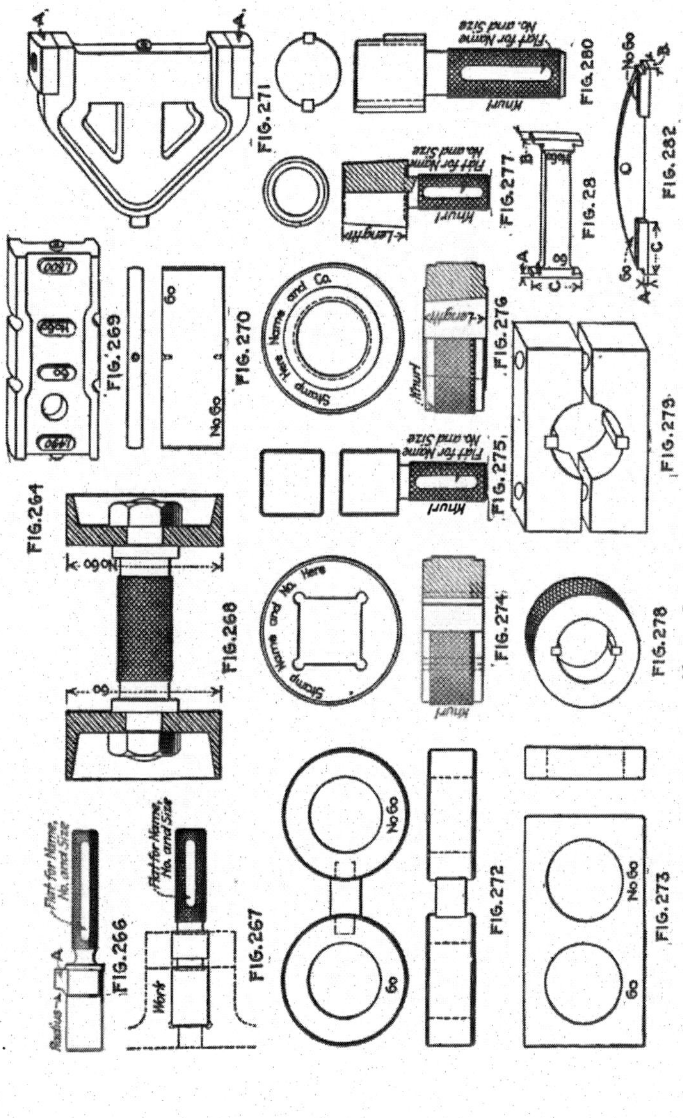

FIGS 266 TO 282.—ANOTHER GROUP OF GAGES.

The sizes from ¼ to 3 ins. will be covered by Figs. 257 and 258, while Fig. 256 will successfully cover the larger requirements. These gages can be made from flat stock as shown in Fig. 258.

PLUG GAGES

For producing holes smaller than can be measured by an inside micrometer, a plug gage is absolutely necessary. An operator who is not provided with such an article by the firm generally takes the firm's time to make one for himself, of which no record is kept, and which is lost when the job is completed. The material of which plug gages should be made is an open question. Some firms make their stock of plug gages of machine steel, case hardened, while others, in an endeavor to save expense, prefer to use a high-carbon steel.

When the gage is high-carbon steel the handles are sometimes made of machine steel. Two methods of doing this are shown in Figs. 259 and 260, but it is doubtful if any advantage is secured by building them up in this way. The disadvantage of Fig. 259 is that the hardened shell is sometimes broken when forcing it on the handles or shank; while in Fig. 260, so little material is really saved that the design becomes more expensive to manufacture, and what is saved on material is more than lost in labor.

STANDARD PLUG GAGES

Standard plug gages such as are shown in Table 22 are used for gaging holes when no limits are called for. The most suitable proportions for general use will be found in the table. For special purposes such as gaging extra long holes the length *C* should be a little longer than the hole to be gaged. A nicely finished and nurled gage demands the workman's respect, and he is much more likely to take care of it than if no trouble or care were taken in its design and finish.

When the size of the plug gage gets beyond 3 in., the question of weight becomes an important factor and calls for alteration in the design. Figs. 261 to 264 show several methods of meeting this condition. A gage grip or handle is shown in Fig. 265. The dimensions shown work out successfully when this is used on large plug gages such as are shown in Figs. 261 to 263. It should be made of cold-drawn steel, and does not require hardening.

A very suitable plug gage for testing the radius at the mouth of a hole is shown in Fig. 266. The body of the gage is a plain plug with a splined hole into which is inserted the flat part *A*, carrying the correct radius. This is secured in position by dowel pins.

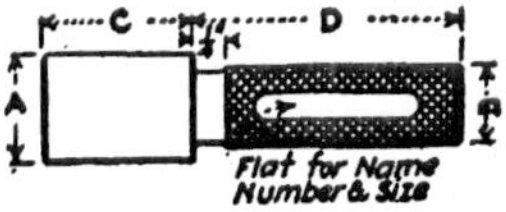

A	B	C	D
1/4	3/16	7/8	2
3/8	5/16	1	2 1/4
1/2	7/16	1 1/8	2 1/2
5/8	9/16	1 1/4	2 3/4
3/4	[illegible]	1 3/8	3
7/8	[illegible]	[illegible]	3 1/8
1	[illegible]	[illegible]	3 1/4
1 1/8	7/8	[illegible]	3 1/4
1 1/4	1	[illegible]	3 3/8
1 3/8	1 1/8	1 7/8	3 3/8
1 1/2	[illegible]	1 7/8	3 3/8
1 5/8	[illegible]	2	3 1/2
1 3/4	[illegible]	2	3 1/2
1 7/8	[illegible]	2 1/8	3 1/2
2	[illegible]	2 1/8	3 1/2
2 1/8	[illegible]	2 1/8	3 1/2
2 1/4	[illegible]	2 1/4	3 5/8
2 3/8	[illegible]	2 1/4	3 5/8
2 1/2	[illegible]	2 1/4	3 5/8
2 5/8	[illegible]	2 1/4	3 5/8
2 3/4	[illegible]	2 1/2	3 5/8
2 7/8	[illegible]	2 1/2	3 5/8
3	[illegible]	2 1/2	3 5/8

All dimensions in inches.
TABLE 22.—PLUG GAGES.

A point to be remembered when designing or making a plug gage is illustrated in Fig. 267. The plug gage acts here as a depth gage as well as a plug gage. The illustration shows it made up to gage the depth of two holes of the same diameter but of different lengths.

LIMIT PLUG GAGES

The proportions for double-ended limit plug gages are given in Table 23. The "go" and "not go" ends are easily distinguishable, being of different lengths. Attention should be drawn to the ends of the gages, which, it will be noticed, are slightly chamfered.

Another form of limit gage is given in Table 24. It is a little more economical to make than the double-ended type and possesses this advantage, that the "go" part lines up the gage to enable the "no

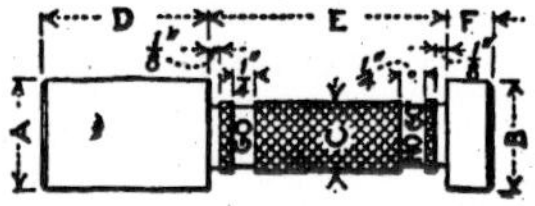

A	B	C	D	E	F
0.249	0.251	$\frac{3}{16}$	$\frac{7}{8}$	$2\frac{1}{2}$	$\frac{1}{4}$
0.374	0.376	$\frac{5}{16}$	1	$2\frac{1}{2}$	$\frac{1}{4}$
0.499	0.501	$\frac{7}{16}$	$1\frac{1}{8}$	$2\frac{1}{2}$	$\frac{1}{4}$
0.624	0.626	$\frac{1}{2}$	$1\frac{1}{4}$	$2\frac{1}{2}$	$\frac{1}{4}$
0.749	0.751	$\frac{5}{8}$	$1\frac{3}{8}$	$2\frac{1}{2}$	$\frac{1}{4}$
0.874	0.876	$\frac{3}{4}$	$1\frac{1}{2}$	$2\frac{1}{2}$	$\frac{3}{8}$
0.999	1.001	$\frac{3}{4}$	$1\frac{5}{8}$	$2\frac{1}{2}$	$\frac{3}{8}$
1.124	1.126	$\frac{3}{4}$	$1\frac{3}{4}$	$2\frac{3}{4}$	$\frac{3}{8}$
1.249	1.251	$\frac{7}{8}$	$1\frac{3}{4}$	$2\frac{3}{4}$	$\frac{3}{8}$
1.374	1.376	$\frac{7}{8}$	$1\frac{7}{8}$	$2\frac{3}{4}$	$\frac{3}{8}$
1.499	1.501	$\frac{7}{8}$	$1\frac{7}{8}$	$2\frac{3}{4}$	$\frac{3}{8}$
1.624	1.626	$\frac{7}{8}$	2	$2\frac{3}{4}$	$\frac{3}{8}$
1.749	1.751	$\frac{7}{8}$	2	$2\frac{3}{4}$	$\frac{3}{8}$
1.874	1.876	1	$2\frac{1}{8}$	$2\frac{3}{4}$	$\frac{3}{8}$
1.999	2.001	1	$2\frac{1}{8}$	3	$\frac{3}{8}$
2.124	2.126	1	$2\frac{1}{8}$	3	$\frac{1}{2}$
2.249	2.251	$1\frac{1}{4}$	$2\frac{1}{4}$	3	$\frac{1}{2}$
2.374	2.376	$1\frac{1}{4}$	$2\frac{1}{4}$	3	$\frac{1}{2}$
2.499	2.501	$1\frac{1}{4}$	$2\frac{1}{4}$	3	$\frac{1}{2}$
2.624	2.626	$1\frac{1}{2}$	$2\frac{1}{4}$	$3\frac{1}{2}$	$\frac{1}{2}$
2.749	2.751	$1\frac{1}{2}$	$2\frac{1}{2}$	$3\frac{1}{2}$	$\frac{1}{2}$
2.874	2.876	$1\frac{1}{2}$	$2\frac{1}{2}$	$3\frac{1}{2}$	$\frac{1}{2}$
2.999	3.001	$1\frac{1}{2}$	$2\frac{1}{2}$	$3\frac{1}{2}$	$\frac{1}{2}$

All dimensions in inches.
TABLE 23.—PLUG LIMIT GAGES.

go" part to be applied to the hole under the best conditions for entering it.

Its use is somewhat restricted by the fact that the end has to pass beyond the length of the hole before the "no go" part can be tried. This confines its practical application to open holes. When the sizes of the limit plug gages get beyond 3 in. the question of weight becomes an important factor and calls for alteration in the design. A method of doing this is shown in Fig. 268.

FLAT PLUG GAGES

A form of gage frequently found in use where a comparatively inexpensive gage is required is shown in Fig. 269. These gages may be made either as double-end limit gages, single-end, maximum and minimum end, and exact size in the center. They are used quite largely as the shop working standards, are lighter than the cylindrical type, and can be laid on the machine or bench without any fear of

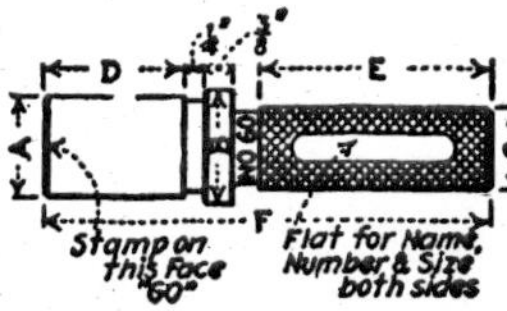

A	B	C	D	E	F
0.249	0.251	$\frac{3}{16}$	$\frac{7}{8}$	2	$3\frac{1}{4}$
0.374	0.376	$\frac{5}{16}$	1	$2\frac{1}{4}$	$4\frac{1}{8}$
0.499	0.501	$\frac{7}{16}$	$1\frac{1}{8}$	$2\frac{1}{2}$	$4\frac{1}{2}$
0.624	0.626	$\frac{1}{2}$	$1\frac{1}{4}$	$2\frac{3}{4}$	$4\frac{7}{8}$
0.749	0.751	$\frac{5}{8}$	$1\frac{3}{8}$	3	$5\frac{1}{4}$
0.874	0.876	$\frac{3}{4}$	$1\frac{1}{2}$	$3\frac{1}{8}$	$5\frac{1}{2}$
0.999	1.001	$\frac{7}{8}$	$1\frac{5}{8}$	$3\frac{1}{4}$	$5\frac{3}{4}$
1.124	1.126	$\frac{7}{8}$	$1\frac{3}{4}$	$3\frac{1}{4}$	$5\frac{7}{8}$
1.249	1.251	1	$1\frac{3}{4}$	$3\frac{3}{8}$	6
1.374	1.376	$1\frac{1}{8}$	$1\frac{7}{8}$	$3\frac{3}{8}$	$6\frac{1}{8}$
1.499	1.501	$1\frac{1}{4}$	$1\frac{7}{8}$	$3\frac{3}{8}$	$6\frac{1}{8}$
1.624	1.626	$1\frac{3}{8}$	2	$3\frac{1}{2}$	$6\frac{3}{8}$
1.749	1.751	$1\frac{1}{2}$	2	$3\frac{1}{2}$	$6\frac{3}{8}$
1.874	1.876	$1\frac{1}{2}$	$2\frac{1}{8}$	$3\frac{1}{2}$	$6\frac{1}{2}$
1.999	2.001	$1\frac{1}{2}$	$2\frac{1}{8}$	$3\frac{1}{2}$	$6\frac{1}{2}$
2.124	2.126	$1\frac{1}{2}$	$2\frac{1}{8}$	$3\frac{1}{2}$	$6\frac{1}{2}$
2.249	2.251	$1\frac{5}{8}$	$2\frac{1}{4}$	$3\frac{5}{8}$	$6\frac{3}{4}$
2.374	2.376	$1\frac{5}{8}$	$2\frac{1}{4}$	$3\frac{5}{8}$	$6\frac{3}{4}$
2.499	2.501	$1\frac{3}{4}$	$2\frac{1}{4}$	$3\frac{5}{8}$	$6\frac{3}{4}$
2.624	2.626	$1\frac{3}{4}$	$2\frac{1}{4}$	$3\frac{5}{8}$	$6\frac{3}{4}$
2.749	2.751	$1\frac{3}{4}$	$2\frac{1}{2}$	$3\frac{5}{8}$	7
2.874	2.876	$1\frac{3}{4}$	$2\frac{1}{2}$	$3\frac{5}{8}$	7
2.999	3.001	$1\frac{3}{4}$	$2\frac{1}{2}$	$3\frac{5}{8}$	7

All dimensions in inches.

TABLE 24.—PLUG LIMIT GAGES.

their rolling about. The wearing surfaces, however, are much reduced.

As a limit gage for large diameter holes, the gage shown in Fig. 270 is ideal and inexpensive. It is easily handled and is not as liable to damage as a pin gage, while its wearing surface is considerably larger.

A built-up flat plug gage is illustrated in Fig. 271. The body can be made of cast iron or aluminum, while the removable pads *A* should be hardened tool steel. These pads are easily removed at small cost and by making different sized pads, the same gage body can be used for a number of different diameters or sizes.

RING GAGES

Suitable proportions for ring gages are presented in Table 25. When making these gages it is advisable to leave a projecting shoulder around the hole, as shown by the dotted lines in the illustration. This shoulder can be removed in the final operation. This cuts off

A	B	C	A	B	C
$\frac{1}{4}$	1	$\frac{1}{4}$	$1\frac{3}{4}$	$2\frac{7}{8}$	$1\frac{3}{8}$
$\frac{3}{8}$	1	$\frac{1}{2}$	$1\frac{7}{8}$	$2\frac{7}{8}$	$1\frac{3}{8}$
$\frac{1}{2}$	1	$\frac{5}{8}$	2	3	$1\frac{1}{2}$
$\frac{5}{8}$	$1\frac{1}{4}$	$\frac{3}{4}$	$2\frac{1}{8}$	$3\frac{1}{8}$	$1\frac{1}{2}$
$\frac{3}{4}$	$\frac{1}{2}$	$\frac{7}{8}$	$2\frac{1}{4}$	$3\frac{1}{4}$	$1\frac{5}{8}$
$\frac{7}{8}$	$1\frac{3}{4}$	1	$2\frac{3}{8}$	$3\frac{1}{2}$	$1\frac{5}{8}$
1	2	1	$2\frac{1}{2}$	$3\frac{3}{4}$	$1\frac{3}{4}$
$1\frac{1}{8}$	$2\frac{1}{4}$	$1\frac{1}{8}$	$2\frac{5}{8}$	$3\frac{7}{8}$	$1\frac{3}{4}$
$1\frac{1}{4}$	$2\frac{1}{2}$	$1\frac{1}{8}$	$2\frac{3}{4}$	4	$1\frac{7}{8}$
$1\frac{3}{8}$	$2\frac{1}{2}$	$1\frac{1}{4}$	$2\frac{7}{8}$	$4\frac{1}{2}$	$1\frac{7}{8}$
$1\frac{1}{2}$	$2\frac{1}{2}$	$1\frac{1}{4}$	3	$4\frac{1}{2}$	$1\frac{7}{8}$
$1\frac{5}{8}$	$2\frac{3}{4}$	$1\frac{3}{8}$			

All dimensions in inches.
TABLE 25.—RING GAGES.

that part of the hole that has become bell-mouthed, due to the grinding wheel cutting more freely at the end.

Two methods of making limit ring gages are shown in Figs. 272 and 273. The built-up gage, Fig. 272, is the easiest to make, but the solid type, Fig. 273, would stand rougher treatment in the shop.

Plug and ring gages for square holes are shown in Figs. 274 and 275. The female part has the corners cut away, and this serves the purpose of allowing clearance for grinding the inside and also does away with the sharp corner which might have the result of starting a crack in the hardening.

TAPER GAGES

A separate set of taper gages should always be provided for turning and grinding sizes. The gage should be made so that the large

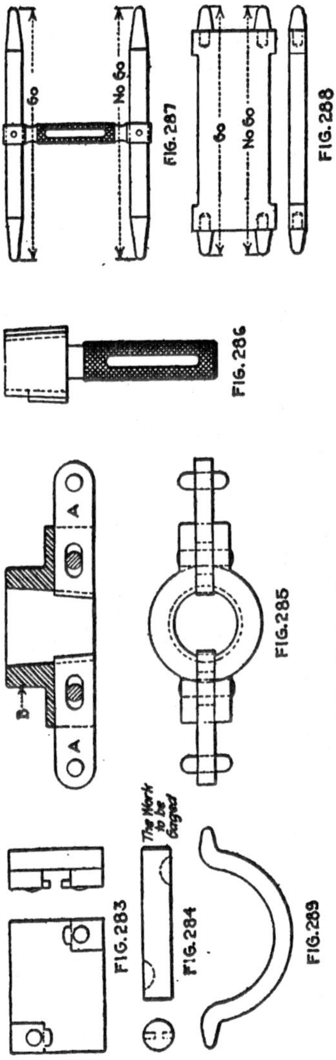

FIGS. 283 TO 289.—KEYWAY AND PIN GAGES.

end of the plug and the small end of the ring come flush with the end of the cone portions, as shown in Figs. 276 and 277. In some shops the same gage is used for the grinding and turning operation of coned surfaces, with instructions to the lathe operator to leave the plug out so far to allow for grinding. Such a practice is fatal to good production; he either leaves too little for grinding, or else too much. Referring again to Fig. 277, it will be seen that the ends are recessed. This is done to allow the gage to be ground on the ends to the correct length of the tapered hole which it is to gage.

The female member, Fig. 276, is treated in the same way, the ends of this gage being ground to the exact length after the bore is finished. It will be observed from the engraving that a shoulder is left for this grinding. This also prevents the marking on the ends from being interfered with.

KEYWAY GAGES

A ring gage for testing the proper location of keyways is illustrated in Fig. 278. Another type, which can be used when it is necessary to apply the gage to a keyway that does not run through to the end of the shaft, is shown in Fig. 279.

The accompanying male plug is seen in Fig. 280. It should be noted in all cases where there are two keys, that only one of them should extend the whole length of the gage. This enables the size of the keyway to be tested independent of its location.

Three styles of limit keyway gages are shown in Figs. 281 and 282 and Table 26. When gaging flat keyways in shafts, Fig. 281 is used and Fig. 282 is used for keyways and holes; *A* represents the depth, *B* the width and *C* the length of the keyway. Table 26 gives suitable proportions for making up a set of limits for Woodruff keyway gages.

An inspector's keyway gage is shown in Fig. 283. It is very good for gaging such work as shown in Fig. 284. The keyways, it will be noticed, are on opposite sides of the shaft.

One method of making female gages for gaging the keyways in a tapered shaft is shown in Fig. 285. With this style of gage, one keyway can be gaged at a time. In a good many cases this gage is superior to the stationary type of keyway gage which is so often made. The keys *A* should be a sliding fit and have no perceptible shake in the body *B*. All parts should be hardened and ground. The gage, Fig. 286, is similar to Fig. 280, except that it is used for gaging the keyways in a tapering hole.

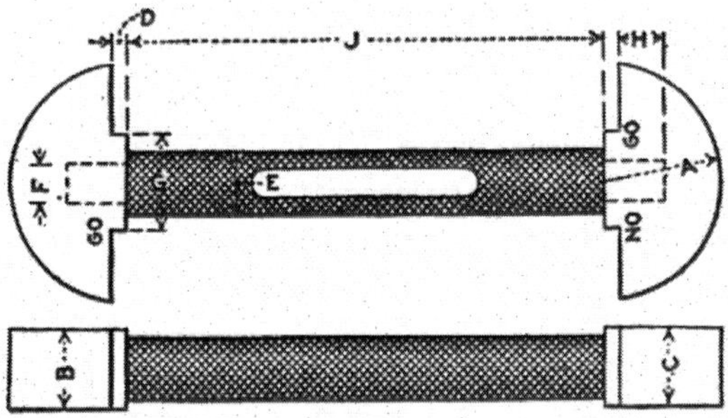

For Key No.	A	B	C	D	E	F	G	H	J
1	$\frac{1}{4}$	$\frac{1}{16}$	$\frac{1}{16}$—002	$\frac{3}{64}$	Use Flat Stock				$1\frac{1}{2}$
2	$\frac{1}{4}$	$\frac{3}{32}$	$\frac{3}{32}$—002	$\frac{3}{64}$					$1\frac{1}{2}$
3	$\frac{1}{4}$	$\frac{1}{8}$	$\frac{1}{8}$—002	$\frac{3}{64}$					$1\frac{1}{2}$
4	$\frac{5}{16}$	$\frac{3}{32}$	$\frac{3}{32}$—002	$\frac{1}{16}$					$1\frac{1}{2}$
5	$\frac{5}{16}$	$\frac{1}{8}$	$\frac{1}{8}$—002	$\frac{1}{16}$					$1\frac{1}{2}$
6	$\frac{5}{16}$	$\frac{5}{32}$	$\frac{5}{32}$—002	$\frac{1}{16}$					$1\frac{1}{2}$
7	$\frac{3}{8}$	$\frac{1}{8}$	$\frac{1}{8}$—002	$\frac{1}{16}$					$1\frac{1}{2}$
8	$\frac{3}{8}$	$\frac{5}{32}$	$\frac{5}{32}$—002	$\frac{1}{16}$					$\frac{1}{2}$
9	$\frac{3}{8}$	$\frac{3}{16}$	$\frac{3}{16}$—002	$\frac{1}{16}$	$\frac{3}{16}$	$\frac{1}{8}$	$\frac{5}{16}$	$\frac{1}{4}$	$1\frac{1}{2}$
10	$\frac{7}{16}$	$\frac{5}{32}$	$\frac{5}{32}$—002	$\frac{1}{16}$	$\frac{3}{16}$	$\frac{5}{32}$	$\frac{5}{16}$	$\frac{1}{4}$	$1\frac{3}{4}$
11	$\frac{7}{16}$	$\frac{3}{16}$	$\frac{3}{16}$—002	$\frac{1}{16}$	$\frac{3}{16}$	$\frac{1}{8}$	$\frac{5}{16}$	$\frac{1}{4}$	$1\frac{3}{4}$
12	$\frac{7}{16}$	$\frac{7}{32}$	$\frac{7}{32}$—002	$\frac{1}{16}$	$\frac{3}{16}$	$\frac{1}{8}$	$\frac{5}{16}$	$\frac{1}{4}$	$1\frac{3}{4}$
A	$\frac{7}{16}$	$\frac{1}{4}$	$\frac{1}{4}$—002	$\frac{1}{16}$	$\frac{3}{16}$	$\frac{1}{8}$	$\frac{5}{16}$	$\frac{1}{4}$	$1\frac{3}{4}$
13	$\frac{1}{2}$	$\frac{3}{16}$	$\frac{3}{16}$—002	$\frac{1}{16}$	$\frac{3}{16}$	$\frac{1}{8}$	$\frac{5}{16}$	$\frac{1}{4}$	2
14	$\frac{1}{2}$	$\frac{7}{32}$	$\frac{7}{32}$—002	$\frac{1}{16}$	$\frac{3}{16}$	$\frac{1}{8}$	$\frac{5}{16}$	$\frac{1}{4}$	2
15	$\frac{1}{2}$	$\frac{1}{4}$	$\frac{1}{4}$—002	$\frac{1}{16}$	$\frac{1}{4}$	$\frac{1}{8}$	$\frac{5}{16}$	$\frac{1}{4}$	2
B	$\frac{1}{2}$	$\frac{5}{16}$	$\frac{5}{16}$—002	$\frac{1}{16}$	$\frac{1}{4}$	$\frac{3}{16}$	$\frac{3}{8}$	$\frac{1}{4}$	2
16	$\frac{9}{16}$	$\frac{3}{16}$	$\frac{3}{16}$—002	$\frac{5}{64}$	$\frac{1}{4}$	$\frac{1}{8}$	$\frac{3}{8}$	$\frac{5}{16}$	2
17	$\frac{9}{16}$	$\frac{7}{32}$	$\frac{7}{32}$—002	$\frac{5}{64}$	$\frac{1}{4}$	$\frac{5}{32}$	$\frac{3}{8}$	$\frac{5}{16}$	2
18	$\frac{9}{16}$	$\frac{1}{4}$	$\frac{1}{4}$—002	$\frac{5}{64}$	$\frac{1}{4}$	$\frac{5}{32}$	$\frac{3}{8}$	$\frac{5}{16}$	2
C	$\frac{9}{16}$	$\frac{5}{16}$	$\frac{5}{16}$—002	$\frac{5}{64}$	$\frac{1}{4}$	$\frac{3}{16}$	$\frac{3}{8}$	$\frac{5}{16}$	2
19	$\frac{5}{8}$	$\frac{3}{16}$	$\frac{3}{16}$—002	$\frac{5}{64}$	$\frac{1}{4}$	$\frac{1}{8}$	$\frac{3}{8}$	$\frac{5}{16}$	$2\frac{1}{4}$
20	$\frac{5}{8}$	$\frac{7}{32}$	$\frac{7}{32}$—002	$\frac{5}{64}$	$\frac{1}{4}$	$\frac{5}{32}$	$\frac{3}{8}$	$\frac{5}{16}$	$2\frac{1}{4}$
21	$\frac{5}{8}$	$\frac{1}{4}$	$\frac{1}{4}$—002	$\frac{5}{64}$	$\frac{1}{4}$	$\frac{5}{32}$	$\frac{3}{8}$	$\frac{3}{8}$	$2\frac{1}{4}$
D	$\frac{5}{8}$	$\frac{5}{16}$	$\frac{5}{16}$—002	$\frac{5}{64}$	$\frac{5}{16}$	$\frac{3}{16}$	$\frac{7}{16}$	$\frac{3}{8}$	$2\frac{1}{4}$
E	$\frac{5}{8}$	$\frac{3}{8}$	$\frac{3}{8}$—002	$\frac{5}{64}$	$\frac{5}{16}$	$\frac{1}{4}$	$\frac{7}{16}$	$\frac{3}{8}$	$2\frac{1}{4}$
22	$\frac{11}{16}$	$\frac{1}{4}$	$\frac{1}{4}$—002	$\frac{3}{32}$	$\frac{5}{16}$	$\frac{5}{32}$	$\frac{7}{16}$	$\frac{3}{8}$	$2\frac{1}{2}$
23	$\frac{11}{16}$	$\frac{5}{16}$	$\frac{5}{16}$—002	$\frac{3}{32}$	$\frac{5}{16}$	$\frac{3}{16}$	$\frac{7}{16}$	$\frac{3}{8}$	$2\frac{1}{2}$
F	$\frac{11}{16}$	$\frac{3}{8}$	$\frac{3}{8}$—002	$\frac{3}{32}$	$\frac{5}{16}$	$\frac{1}{4}$	$\frac{7}{16}$	$\frac{3}{8}$	$2\frac{1}{2}$
24	$\frac{3}{4}$	$\frac{1}{4}$	$\frac{1}{4}$—002	$\frac{7}{64}$	$\frac{5}{16}$	$\frac{5}{32}$	$\frac{1}{2}$	$\frac{3}{8}$	$2\frac{1}{2}$
25	$\frac{3}{4}$	$\frac{5}{16}$	$\frac{5}{16}$—002	$\frac{7}{64}$	$\frac{5}{16}$	$\frac{1}{4}$	$\frac{1}{2}$	$\frac{3}{8}$	$2\frac{1}{2}$
G	$\frac{3}{4}$	$\frac{3}{8}$	$\frac{3}{8}$—002	$\frac{7}{64}$	$\frac{5}{16}$	$\frac{1}{4}$	$\frac{1}{2}$	$\frac{3}{8}$	$2\frac{1}{2}$

All dimensions in inches.
TABLE 26.—WOODRUFF KEY GAGES.

A	B	A	B
3	$\frac{5}{16}$	$7\frac{3}{4}$	$\frac{7}{16}$
$3\frac{1}{4}$	$\frac{5}{16}$	8	$\frac{7}{16}$
$3\frac{1}{2}$	$\frac{5}{16}$	$8\frac{1}{4}$	$\frac{7}{16}$
$3\frac{3}{4}$	$\frac{5}{16}$	$8\frac{1}{2}$	$\frac{7}{16}$
4	$\frac{5}{16}$	$8\frac{3}{4}$	$\frac{7}{16}$
$4\frac{1}{4}$	$\frac{5}{16}$	9	$\frac{1}{2}$
$4\frac{1}{2}$	$\frac{5}{16}$	$9\frac{1}{4}$	$\frac{1}{2}$
$4\frac{3}{4}$	$\frac{5}{16}$	$9\frac{1}{2}$	$\frac{1}{2}$
5	$\frac{3}{8}$	$9\frac{3}{4}$	$\frac{1}{2}$
$5\frac{1}{4}$	$\frac{3}{8}$	10	$\frac{1}{2}$
$5\frac{1}{2}$	$\frac{3}{8}$	$10\frac{1}{4}$	$\frac{1}{2}$
$5\frac{3}{4}$	$\frac{3}{8}$	$10\frac{1}{2}$	$\frac{1}{2}$
6	$\frac{3}{8}$	$10\frac{3}{4}$	$\frac{1}{2}$
$6\frac{1}{4}$	$\frac{3}{8}$	11	$\frac{9}{16}$
$6\frac{1}{2}$	$\frac{3}{8}$	$11\frac{1}{4}$	$\frac{9}{16}$
$6\frac{3}{4}$	$\frac{3}{8}$	$11\frac{1}{2}$	$\frac{9}{16}$
7	$\frac{7}{16}$	$11\frac{3}{4}$	$\frac{9}{16}$
$7\frac{1}{4}$	$\frac{7}{16}$	12	$\frac{9}{16}$
$7\frac{1}{2}$	$\frac{7}{16}$		

All dimensions in inches.
TABLE 27.—PIN GAGES.

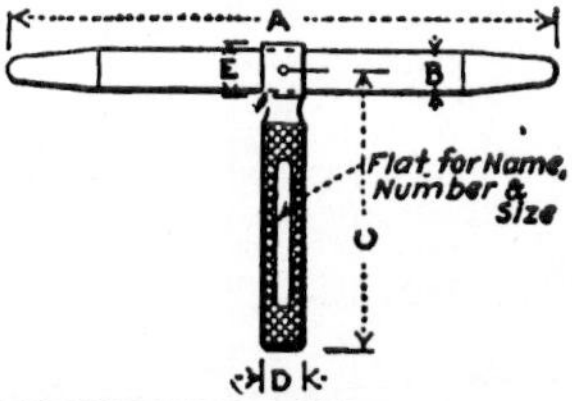

A	B	C	D	E	A	B	C	D	E
2	$\frac{1}{4}$	$1\frac{3}{4}$	$\frac{1}{4}$	$\frac{3}{8}$	$3\frac{5}{8}$	$\frac{5}{16}$	$2\frac{1}{8}$	$\frac{5}{16}$	$\frac{7}{16}$
$2\frac{1}{8}$	$\frac{1}{4}$	$1\frac{3}{4}$	$\frac{1}{4}$	$\frac{3}{8}$	$3\frac{3}{4}$	$\frac{5}{16}$	$2\frac{1}{8}$	$\frac{5}{16}$	$\frac{7}{16}$
$2\frac{1}{4}$	$\frac{1}{4}$	$1\frac{3}{4}$	$\frac{1}{4}$	$\frac{3}{8}$	$3\frac{7}{8}$	$\frac{5}{16}$	$2\frac{1}{8}$	$\frac{5}{16}$	$\frac{7}{16}$
$2\frac{3}{8}$	$\frac{1}{4}$	$1\frac{3}{4}$	$\frac{1}{4}$	$\frac{3}{8}$	4	$\frac{5}{16}$	$2\frac{1}{4}$	$\frac{5}{16}$	$\frac{7}{16}$
$2\frac{1}{2}$	$\frac{1}{4}$	$1\frac{3}{4}$	$\frac{1}{4}$	$\frac{3}{8}$	$4\frac{1}{8}$	$\frac{5}{16}$	$2\frac{1}{4}$	$\frac{5}{16}$	$\frac{7}{16}$
$2\frac{5}{8}$	$\frac{1}{4}$	$1\frac{3}{4}$	$\frac{1}{4}$	$\frac{3}{8}$	$4\frac{1}{4}$	$\frac{5}{16}$	$2\frac{1}{4}$	$\frac{5}{16}$	$\frac{7}{16}$
$2\frac{3}{4}$	$\frac{1}{4}$	$1\frac{3}{4}$	$\frac{1}{4}$	$\frac{3}{8}$	$4\frac{3}{8}$	$\frac{5}{16}$	$2\frac{1}{4}$	$\frac{5}{16}$	$\frac{7}{16}$
$2\frac{7}{8}$	$\frac{1}{4}$	$1\frac{3}{4}$	$\frac{1}{4}$	$\frac{3}{8}$	$4\frac{1}{2}$	$\frac{5}{16}$	$2\frac{1}{2}$	$\frac{5}{16}$	$\frac{7}{16}$
3	$\frac{5}{16}$	2	$\frac{5}{16}$	$\frac{7}{16}$	$4\frac{5}{8}$	$\frac{5}{16}$	$2\frac{1}{2}$	$\frac{5}{16}$	$\frac{7}{16}$
$3\frac{1}{8}$	$\frac{5}{16}$	2	$\frac{5}{16}$	$\frac{7}{16}$	$4\frac{3}{4}$	$\frac{5}{16}$	$2\frac{1}{2}$	$\frac{5}{16}$	$\frac{7}{16}$
$3\frac{1}{4}$	$\frac{5}{16}$	2	$\frac{5}{16}$	$\frac{7}{16}$	$4\frac{7}{8}$	$\frac{5}{16}$	$2\frac{1}{2}$	$\frac{5}{16}$	$\frac{7}{16}$
$3\frac{3}{8}$	$\frac{5}{16}$	2	$\frac{5}{16}$	$\frac{7}{16}$	5	$\frac{3}{8}$	$2\frac{1}{2}$	$\frac{3}{8}$	$\frac{1}{2}$
$3\frac{1}{2}$	$\frac{5}{16}$	$2\frac{1}{8}$	$\frac{5}{16}$	$\frac{7}{16}$					

All dimensions in inches.
TABLE 28.—PIN GAGES WITH HANDLES.

PIN GAGES

Holes can be very accurately tested by pin gages, but when used on small holes it is necessary that they should be in competent hands. For large holes they are the only gages that can be used at all. The proportions given in Tables 27 and 28 give a stiff reliable gage without

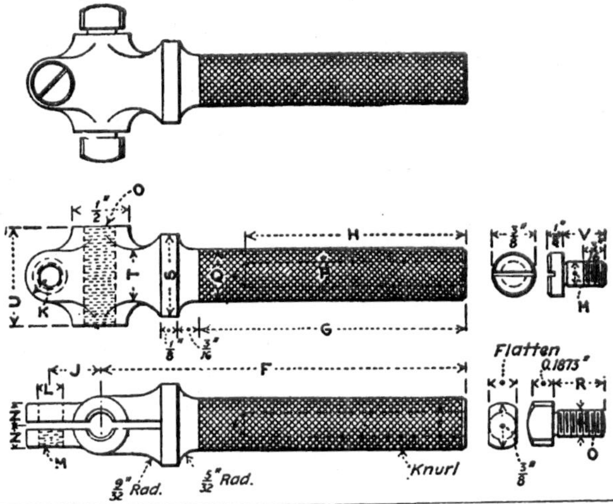

Size.	F	G	H	J	K	L	M	N	O	P	Q	R	S	T	U	V
1 to 1¼	3⅛	2¼	2	7/16	7/32		No. 10–32 THD	11/64	¼–24 THD	¼	7/16	5/16	11/16	7/16	⅝	11/32
1¼ to 1⅝	3⅛	2¼	2	7/16	7/32	13/64	No. 10–32 THD	11/64	¼–24 THD	¼	7/16	7/16	11/16	7/16	⅞	11/32
1⅝ to 2¼	3½	2⅝	2¾	½	¼	17/64	¼–24 THD	7/32	5/16–20 THD	7/16	⅝	⅝	⅞	½	1¼	7/16
2¼ to 3	3½	2⅝	2¾	½	¼	17/64	¼–24 THD	7/32	5/16–20 THD	7/16	⅝	15/16	⅞	½	1⅞	7/16

All dimensions in inches.

TABLE 29.—ADJUSTABLE PIN GAGES.

being unnecessarily cumbrous. The addition of the handle is particularly advisable for smaller sizes or for gaging the bore of holes of any depth. This type of gage adapted to the limit system is shown in Fig. 287.

Fig. 288 is another design which can be used for the same purpose. In the larger sizes of this example the center of the plate is cut out to make it lighter and more easily handled.

A set of pin gages in which a bend is introduced for the purpose of allowing it to be used at some distance in a hole, or applied while the work is on a mandrel, in a manner similar to the gage shown in Fig. 246, is shown in Fig. 289.

A very useful adjustable pin gage is shown in Table 29. The proportions have been found suitable in practice. The adjustable pins are of hardened tool steel, while all other parts are of machine steel. When adjusted to the proper diameter, these pins are locked by tightening the clamping screws. The heads of the pins are ground to the radius of the smallest hole in which they can be used.

CHAPTER XII

EXTERNAL AND INTERNAL THREAD GAGES

External and internal thread gages are as essential as male and female gages which are used for testing bored or turned surfaces. In most of our large factories these thread gages are made up in three sets; namely, reference, inspection and machine. The reference gages

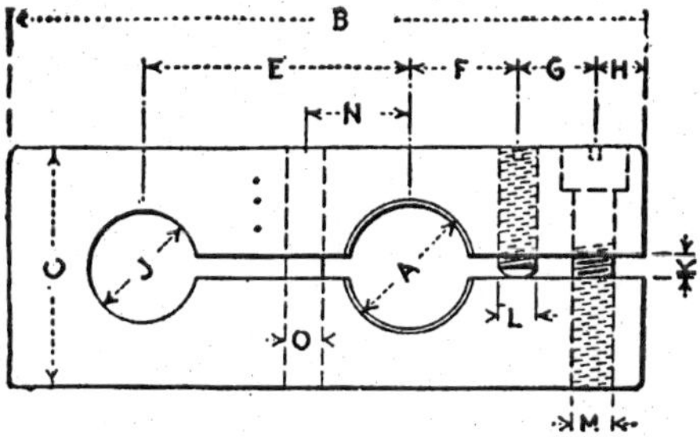

A	B	C	D	E	F	G	H	J	K	L	M	N	O
1/4	2	3/4	5/16	7/8	9/32	1/4	3/16	3/8	1/16	1/8	1/8	1/4	1/8
5/16	2 1/8	3/4	5/16	1	9/32	1/4	3/16	3/8	1/16	1/8	1/8	1/4	1/8
3/8	2 1/4	7/8	3/8	1 1/16	11/32	1/4	3/16	3/8	1/16	1/8	1/8	5/16	1/8
7/16	2 3/8	7/8	3/8	1 1/8	11/32	1/4	3/16	3/8	1/16	1/8	1/8	5/16	1/8
1/2	2 5/8	1	7/16	1 1/4	13/32	5/16	3/16	3/8	1/16	3/16	3/16	13/32	3/16
9/16	2 3/4	1	7/16	1 3/8	13/32	5/16	3/16	3/8	1/16	3/16	3/16	13/32	3/16
5/8	3	1 1/4	1/2	1 1/2	1/2	5/16	3/16	1/2	1/16	3/16	3/16	1/2	7/32
11/16	3 1/8	1 1/4	1/2	1 5/8	1/2	5/16	3/16	1/2	1/16	3/16	3/16	1/2	7/32
3/4	3 3/8	1 3/8	9/16	1 3/4	19/32	5/16	3/16	1/2	3/32	1/4	1/4	9/16	7/32
7/8	3 3/4	1 1/2	9/16	1 7/8	11/16	3/8	1/4	1/2	3/32	1/4	1/4	11/16	1/4
1	4 1/4	1 3/4	5/8	2 3/16	13/16	3/8	5/16	5/8	1/8	1/4	1/4	3/4	1/4
1 1/8	4 3/8	1 7/8	5/8	2 3/16	7/8	3/8	5/16	5/8	1/8	1/4	1/4	3/4	1/4
1 1/4	4 1/2	2	11/16	2 1/4	7/8	3/8	5/16	5/8	1/8	1/4	1/4	7/8	1/4

All dimensions in inches

TABLE 30.—PROPORTIONS FOR EXTERNAL THREAD GAGES 1/4 TO 1 1/4 IN. DIAMETER.

are only used when making up new gages and the manufacturing as well as the inspection departments have their own sets which are exact duplicates of the original reference. This insures interchangeable manufacture of all threaded parts and the reference set of gages

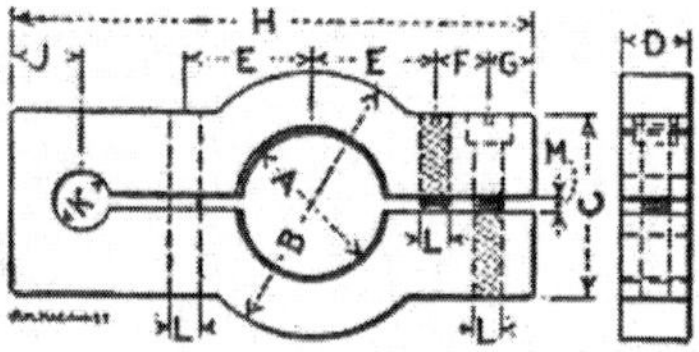

A	B	C	D	E	F	G	H	J	K	L	M
1½	2⅝	1⅝	⅝	1¼	½	⅜	4⅝	⅝	½	¼	⅛
1⅝	2¾	1⅝	⅝	1¼	½	⅜	4¾	⅝	½	¼	⅛
1¾	2⅞	1⅝	⅝	1⅜	½	⅜	5	⅝	½	¼	⅛
1⅞	3	1⅝	¾	1 7/16	½	⅜	5⅛	⅝	½	¼	⅛
2	3⅛	1¾	¾	1½	½	⅜	5½	¾	½	¼	⅛
2⅛	3¼	1¾	¾	1⅝	½	7/16	5⅝	¾	½	¼	⅛
2¼	3⅜	1¾	¾	1 11/16	½	7/16	5⅞	¾	½	¼	⅛
2⅜	3½	2	⅞	1¾	½	7/16	6	11/16	½	¼	⅛
2½	3⅝	2	⅞	1¾	½	7/16	6	11/16	½	¼	⅛

All sizes in inches

TABLE 31.—EXTERNAL THREAD GAGES 1½ TO 2½ IN. DIAMETER.

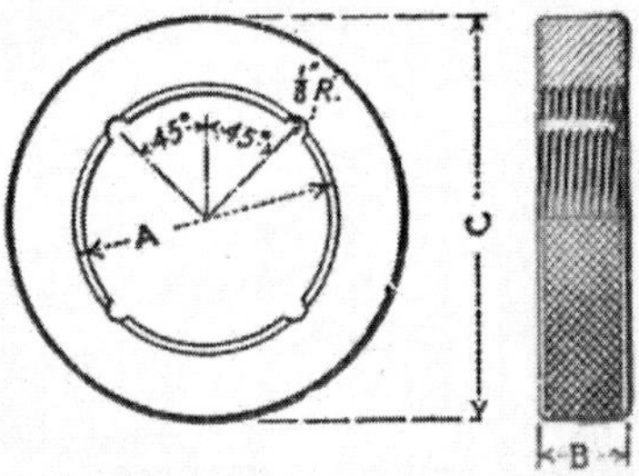

A	B	C	A	B	C	A	B	C
2⅝	⅞	3⅞	3⅜	⅞	4¾	5	⅞	6½
2 11/16	⅞	3⅞	3½	⅞	4⅞	5¼	⅞	6¾
2¾	⅞	4	3⅝	⅞	5	5½	⅞	7
2 13/16	⅞	4	3¾	⅞	5⅛	5¾	⅞	7¼
2⅞	⅞	4⅛	3⅞	⅞	5¼	6	⅞	7½
2 15/16	⅞	4⅛	4	⅞	5⅜			
3	⅞	4¼	4¼	⅞	5⅝			
3⅛	⅞	4⅜	4½	⅞	5⅞			
3¼	⅞	4½	4¾	⅞	6¼			

All sizes in inches

TABLE 32.—EXTERNAL THREAD GAGES 2⅝ TO 3¼ IN. DIAMETER.

constitute a standard, and they always readily settle disputes between the manufacturing and inspection departments.

Table 30 gives suitable proportions for external thread gages from 1/4 to 1 1/4 in. Table 31 shows another form or method of making them from 1 1/2 to 2 1/2 in., and for sizes from 2 5/8 to 6 in. the proportions are

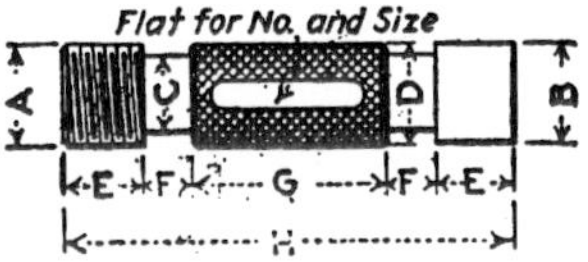

A	B	C	D	E	F	G	H
1/4	Required diameter at bottom of thread	3/16	1/4	1/4	1/4	1 3/4	2 3/4
5/16		1/4	5/16	5/16	1/4	1 3/4	2 7/8
3/8		5/16	3/8	3/8	1/4	1 3/4	3
7/16		3/8	7/16	7/16	5/16	1 5/8	3 1/8
1/2		7/16	1/2	1/2	5/16	1 5/8	3 1/4
9/16		1/2	9/16	1/2	5/16	1 3/4	3 3/8
5/8		1/2	5/8	1/2	3/8	1 3/4	3 1/2
11/16		9/16	5/8	5/8	3/8	1 5/8	3 5/8
3/4		5/8	3/4	5/8	3/8	1 3/4	3 3/4
13/16		11/16	3/4	5/8	7/16	1 3/4	3 7/8
7/8		3/4	7/8	3/4	7/16	1 5/8	4
1		7/8	1	3/4	1/2	1 3/4	4 1/4
1 1/8		7/8	1	3/4	1/2	2	4 1/2
1 1/4		7/8	1	3/4	1/2	2 1/2	5
1 3/8		7/8	1	3/4	1/2	2 1/2	5
1 1/2		7/8	1	3/4	1/2	2 1/2	5
1 5/8		7/8	1	3/4	1/2	2 3/4	5 1/4
1 3/4		7/8	1	3/4	1/2	2 3/4	5 1/4
1 7/8		7/8	1	7/8	1/2	2 1/2	5 1/4
2		1	1 1/8	7/8	1/2	2 3/4	5 1/2
2 1/4		1	1 1/8	7/8	1/2	2 3/4	5 1/2
2 1/2		1	1 1/8	7/8	1/2	2 3/4	5 1/2

All sizes in inches

TABLE 33.—INTERNAL THREAD GAGES 1/4 TO 2 1/2 IN. DIAMETER.

given in Table 32. As a general rule, the external thread gage is fitted to the internal. This really makes the internal gages the practical working standards. The threads are turned and ground to exact size, after which they are hardened, then ground and lapped. In making the internal gages it is always good practice to leave the bottom of the thread sharp to give clearance.

Two methods of designing internal thread gages are shown in Tables 33 and 34. The former gives the necessary dimensions for making them from ¼ to 2½ in., and for all sizes from 2⅝ to 6 in. the latter table can be used.

Figs. 290 to 293 are lapping blocks which, if made of cast iron and split similar to the style of gages shown in Tables 35 and 36, give excel-

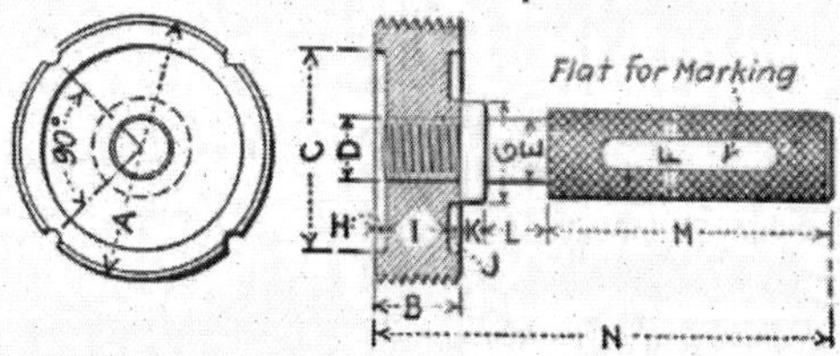

A	B	C	D	E	F	G	H	I	J	K	L	M	N
2⅝	⅞	1⅝	[illegible]-11	[illegible]	⅞	1	[illegible]	[illegible]	[illegible]	[illegible]	[illegible]	3	4⅞
2 11/16	⅞	1 11/16	[illegible]-11	[illegible]	⅞	1	[illegible]	[illegible]	[illegible]	[illegible]	[illegible]	3	4⅞
2¾	⅞	1¾	[illegible]-11	[illegible]	⅞	1	[illegible]	[illegible]	[illegible]	[illegible]	[illegible]	3	4⅞
2 13/16	⅞	1 13/16	[illegible]-11	[illegible]	⅞	1	[illegible]	[illegible]	[illegible]	[illegible]	[illegible]	3	4⅞
2⅞	⅞	1⅞	[illegible]-11	[illegible]	⅞	1	[illegible]	[illegible]	[illegible]	[illegible]	[illegible]	3	4⅞
2 15/16	⅞	1 15/16	[illegible]-11	[illegible]	⅞	1	[illegible]	[illegible]	[illegible]	[illegible]	[illegible]	3	4⅞
3	⅞	2	[illegible]-11	[illegible]	⅞	1	[illegible]	[illegible]	[illegible]	[illegible]	[illegible]	3	4⅞
3⅛	⅞	2⅛	[illegible]-11	[illegible]	⅞	1	[illegible]	[illegible]	[illegible]	[illegible]	[illegible]	3	4⅞
3¼	⅞	2¼	[illegible]-11	[illegible]	⅞	1	[illegible]	[illegible]	[illegible]	[illegible]	[illegible]	3	4⅞
3⅜	⅞	2⅜	[illegible]-11	[illegible]	⅞	1	[illegible]	[illegible]	[illegible]	[illegible]	[illegible]	3	4⅞
3½	⅞	2½	[illegible]-11	[illegible]	⅞	1	[illegible]	[illegible]	[illegible]	[illegible]	[illegible]	3	4⅞
3⅝	⅞	2⅝	[illegible]-11	[illegible]	⅞	1	[illegible]	[illegible]	[illegible]	[illegible]	[illegible]	3	4⅞
3¾	⅞	2¾	[illegible]-11	[illegible]	⅞	1	[illegible]	[illegible]	[illegible]	[illegible]	[illegible]	3	4⅞
3⅞	⅞	2⅞	[illegible]-11	[illegible]	⅞	1	[illegible]	[illegible]	[illegible]	[illegible]	[illegible]	3	4⅞
4	⅞	3	[illegible]-10	[illegible]	1	1¼	[illegible]	[illegible]	[illegible]	[illegible]	[illegible]	3	5
4¼	⅞	3¼	[illegible]-10	[illegible]	1	1¼	[illegible]	[illegible]	[illegible]	[illegible]	[illegible]	3	5
4½	⅞	3½	[illegible]-10	[illegible]	1	1¼	[illegible]	[illegible]	[illegible]	[illegible]	[illegible]	3	5
4¾	⅞	3¾	[illegible]-10	[illegible]	1	1¼	[illegible]	[illegible]	[illegible]	[illegible]	[illegible]	3	5
5	⅞	4	[illegible]-9	[illegible]	1⅛	1⅜	[illegible]	[illegible]	[illegible]	[illegible]	[illegible]	3	5
5¼	⅞	4	[illegible]-9	[illegible]	1⅛	1⅜	[illegible]	[illegible]	[illegible]	[illegible]	[illegible]	3	5
5½	⅞	4⅛	[illegible]-9	[illegible]	1⅛	1⅜	[illegible]	[illegible]	[illegible]	[illegible]	[illegible]	3	5
5¾	⅞	4¼	[illegible]-9	[illegible]	1⅛	1⅜	[illegible]	[illegible]	[illegible]	[illegible]	[illegible]	3	5
6	⅞	4½	[illegible]-9	[illegible]	1⅛	1⅜	[illegible]	[illegible]	[illegible]	[illegible]	[illegible]	3	5

All dimensions in inches.

TABLE 34.—INTERNAL GAGES 2⅝ TO 6 IN. DIAMETER.

lent results for lapping internal thread gages. The sizes given correspond to the table mentioned above.

Another form of female thread gage is shown in Fig. 294. The gage in this case is split at three equal points, adjusted by a screw and locked by a nut. The screw, it will be noticed, is a combination screw and dowel pin.

Fig. 295 represents an adjustable male thread-gage. It consists of a disk split in six equal places, which fits over a tapered plug handle. To adjust this gage it is necessary to grind off the shoulder of the tapered portion an amount sufficient to obtain the correct size. It can then be locked by the washer and set screw. A keyway should

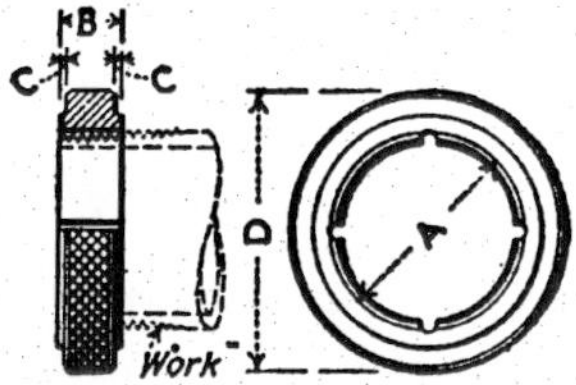

Size of Pipe	Outside Diam. of Pipe or Large End of Ring A	Length of Perfect Thread or Width of Ring B	Length of Shoulder C	Outside Diameter of Ring D
1/8	0.405	0.19	1/32	7/8
1/4	0.540	0.29	1/32	1
3/8	0.675	0.30	1/32	1 1/4
1/2	0.840	0.39	1/32	1 1/2
3/4	1.050	0.40	1/32	1 3/4
1	1.315	0.51	1/16	2
1 1/4	1.660	0.54	1/16	2 5/8
1 1/2	1.900	0.55	1/16	3 1/8
2	2.375	0.58	1/16	3 1/2
2 1/2	2.875	0.89	1/8	4 1/8
3	3.500	0.95	1/8	4 7/8
3 1/2	4.000	1.00	1/8	5 3/8

All sizes in inches.

TABLE 35.—RING GAGES FOR PIPE THREADS.

be placed in the tapered shank to prevent the disk from turning when the gage is in use.

A reference thread gage is shown in Fig. 296. This style of gage is used only as a final reference gage. Calipers are used in measuring from these gages, and the working gages as a general rule are referred to them in case of dispute. They are not hardened.

A style of caliper thread gage is shown in Figs. 297 and 298. The working points of this gage are hardened tool-steel pins, which can be removed and reground or new ones made to replace broken or worn points. By making these points adjustable one gage can be made so as to take the place of three or four of the ordinary type

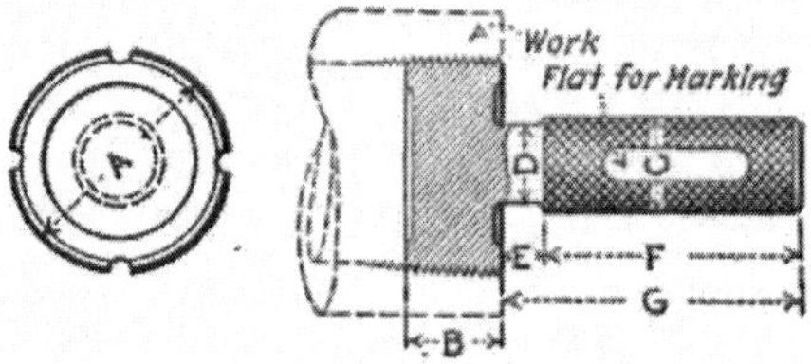

Size	Outside Diam. of Perfect Threads or Large Diam. of Plug A	Length of Plug B	Diam. of Handle C	Diam. of Shoulder of Handle D	Length of Shoulder E	Length of Handle F	Over-all Length of Handle G
1/8	0.405	Three threads longer than length of ring	7/16	3/8	5/16	1 3/8	1 15/16
1/4	0.540		9/16	1/2	5/16	1 1/4	2 1/16
3/8	0.675		5/8	9/16	3/8	1 [illegible]	2
1/2	0.840		3/4	[illegible]	7/16	1 [illegible]	2 1/16
3/4	1.050		1	[illegible]	1/2	1 1/2	2 1/4
1	1.315		1	[illegible]	1/2	2 1/8	3
1 1/4	1.660		1	[illegible]	1/2	2 1/4	3 1/4
1 1/2	1.900		1	[illegible]	1/2	2 1/4	3 1/4
2	2.375		1 1/8	[illegible]	5/8	2 [illegible]	3 [illegible]
2 1/2	2.875		1 [illegible]	[illegible]	5/8	3	3 5/8
3	3.50		1 [illegible]	[illegible]	3/4	3	3 3/4
3 1/2	4.00		1 [illegible]	7/8	7/8	3	3 7/8

All sizes in inches.

TABLE 36.—PLUG GAGES FOR PIPE THREADS.

Their use, however, would necessitate having a set of master-gage rods to adjust them to. Fig. 297 is used for work from 3 to 10 in. in diameter, while Fig. 298 takes care of all sizes under 3 in.

LIMIT GAGES

Limit gages for screw threads can be used to advantage in the manufacturing and inspection departments. With the ordinary male and female thread gage it is next to impossible to judge the limits governing the thread which is to pass inspection. The gage shown in Fig. 299 is similar to an ordinary limit plug gage.

The short end acts as the "no go" and the long end the "go." This style of gage will keep the tapped holes within such limits as is desired. Figs. 300 and 301 are similar to Figs. 297 and 298, and govern the same sizes. Their qualifications are similar and they are made up the same except with the addition of the extra set of points which makes them limit gages.

Another form of a limit external thread gage is shown in Fig. 302. This is a combination of pitch as well as pitch-diameter checking gage.

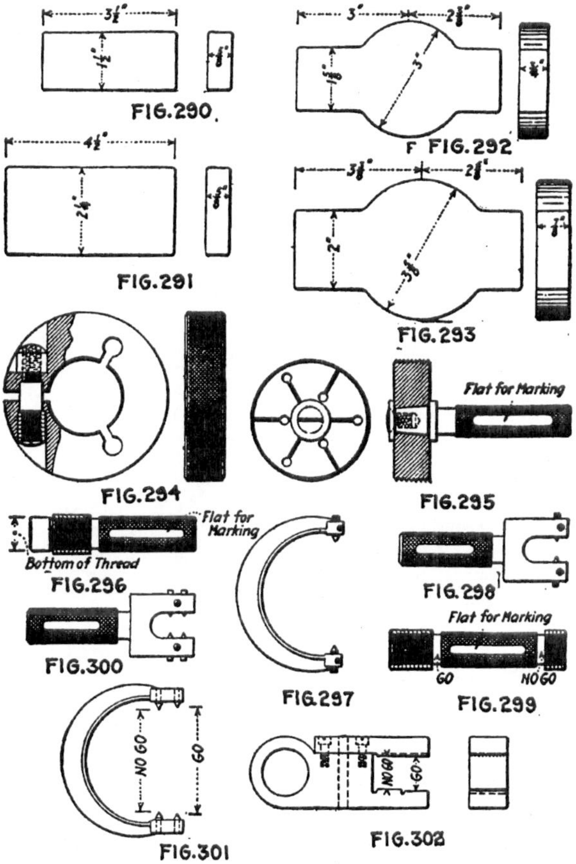

FIGS. 290 TO 302.—EXTERNAL AND INTERNAL THREAD GAGES.

The "go" end of the jaw differs from the "no go" by the amount equal to the tolerance to be allowed on the pitch diameter of the screw. The upper jaw is formed with parallel grooves, which conform to the style, size, etc., of thread to be gaged.

Screw-machine work can be gaged in one-tenth the time it requires to screw a ring gage on and off. In the inspection department, where a large amount of this class of work has to be examined or checked, limits are absolutely necessary, and it is claimed that 30 per cent. of the time ordinarily spent in inspecting such work can be saved with gages of this description.

SCREW-PITCH AND PIPE-THREAD GAGES

These gages are very useful in furnishing a quick way to tell the pitch of a screw and accuracy of the thread. They will very readily show any slight variation from the true pitch, thus making them desirable for the use of those who wish to determine the change of the pitch resulting from the hardening of taps, screws, etc. They should be made very accurately and about 4 in. long.

The fact that wrought-iron pipe is made by different manufacturers from those making the fittings has led manufacturers to adopt one standard which is the "Briggs." These gages have a uniform taper of 3/4 in. to the foot or an inclination to the axis of 1 in 32. The ring gage, Table 6, should be made of a thickness equal to the length of the threads on the end of the pipe, which are perfect at both top and bottom and are used to test pipe ends by screwing them on as far as possible by hand until the end of the pipe is exactly flush with the small end of the gage. This is known as a working gage.

The plug gages, Table 7, should be made of a length three threads greater than the ring gage. The large end being the same size it is screwed in until it comes flush with the large end of the work (see illustration). Thus it will be noted that the fitting is gaged three threads deeper than the pipe end will screw in by hand, this amount being the allowance for screwing down with the wrench to make a tight joint.

Whenever possible, tool steel should be used in the construction of the smaller sizes of thread gages. For the large sizes a good grade of low-carbon steel is the only suitable material that can be used. In both cases they should be hardened by a process which will have the pitch and angle of the thread correct. A highly finished thread gives the best wearing qualities.

CHAPTER XIII

MISCELLANEOUS MANUFACTURING GAGES

Very useful and inexpensive gages known as target gages are shown in Figs. 303 to 307. These gages are used mostly in the milling and inspection departments. In the milling department they are very useful in lining up the work from a given number of bosses or from the profile, etc., of the part to be machined before the milling operation. Figs. 303 to 305 show the stationary type, Fig. 306 the adjustable type, and Fig. 307 an inspector's target gage, used for checking up rough castings or forgings.

The gage shown would be found suitable for checking up connecting-rod forgings such as are used in the automobile industry. The target part of these gages should be made of $\frac{1}{16}$- to $\frac{1}{8}$-in. sheet steel. They then can be riveted, screwed or soldered to the connecting member. It is unnecessary to harden any part of these gages except in such cases where they require a support or base.

CHECKING GAGES

Checking gages are generally called for when it is necessary to check up parts after they have been finished. Fig. 308 shows a solid type, and Fig. 309 a hinged gage. Fig. 310 illustrates a type of checking gage used for gaging the profile of the work, Fig. 311, with its relation to the hole as well as keyway. Figs. 312 and 314 are similar; they are very useful and great time savers, particularly when it is necessary to check up such work as is shown in Figs. 313 and 315.

In both cases it is necessary to gage the bolt holes with their relation to the center hole; the gages are, therefore, a combination of plug gages. The pins *A* are the same diameter as the studs over which the work is to go in the assembling operation, and these gages will readily inform the inspector or foreman of the relative accuracy of the work.

Three more similar types of gages, known as slot and ring checking gages, are shown in Figs. 316 to 318. Fig. 316 illustrates a method of making a gage for testing grooves or making slots. The dimension *A* represents the width and *B* the depth of the slot to be gaged. Fig.

317 is the female gage which is to test the width of a ring, tongue, etc. The groove *A* represents the width. Fig. 318 is the same style gage shown in Fig. 317, except that it is made as a limit gage; one groove acts as the "go" and the other, "no go."

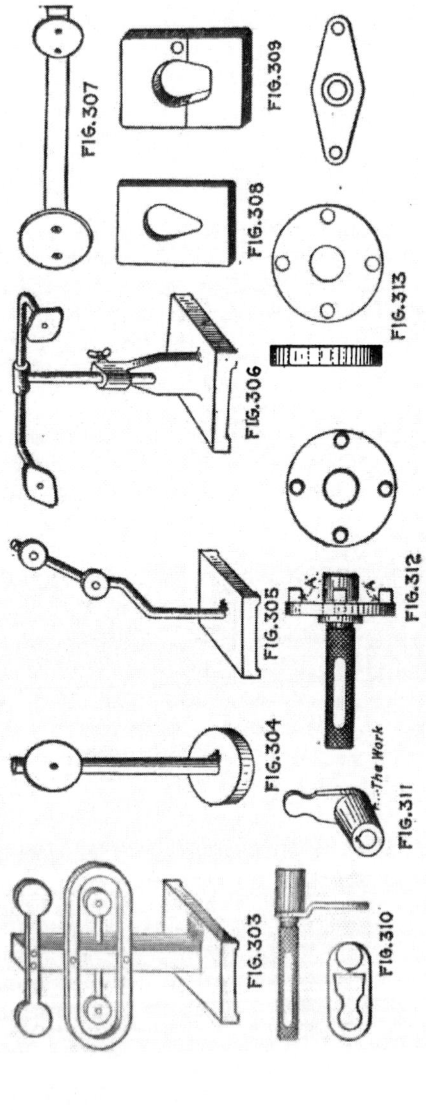

FIGS. 303 TO 314.—TARGET, CHECKING AND OTHER GAGES.

HEIGHT GAGES

Two forms of height gages are shown in Figs. 319 and 320. Ordinarily, gages of this description are made adjustable and graduated; nevertheless there are times when this is unnecessary, and the gages shown illustrate an inexpensive method of designing or making them.

RADIUS GAGES

One of the most useful articles in the gage line is a male and female radius checking gage, such as shown in Figs. 321 and 322. Fig. 321 represents the female and Fig. 322 the male gage. By making or designing them as shown in the illustrations, each corner representing a different radius, a small number of them will cover all the requirements of the shop. In some instances, the ring gage is made to take care of the radius as well as the diameter of the work, and in such cases they can be made as shown in Fig. 323.

FILLISTER AND FLAT-HEAD GAGES

The proportions for gages used in gaging the drill and counterbore of fillister-head screws is shown in Table 37, *D* representing the body of the fillister-head screw, *A* the diameter of the head or size of counterbore to be used, and *C* the length of the head or depth of the counterbore.

Table 38 represents suitable proportions for gages used in gaging the size drill and countersink for flat-head machine screws. The diameter of the screw is represented by *D*; *A* is the diameter and *C* the length of the head.

PROFILE GAGES

To assist the maker or designer in producing a set of satisfactory profile gages for a certain component part, an example is given in Fig. 324. To manufacture this it will be necessary to perform two operations; therefore it is advisable to construct the profile gage in such a manner as to prevent error as well as to attain the highest efficiency in handling.

In this case the work is done on a screw machine and made from bar stock. The gage for this, the first operation, is shown in Fig. 325. It is made to check up the first three shoulders, also the length over all, and may be called a screw-machine gage.

The gage, Fig. 326, for the second operation, which may be done on a lathe or hand-screw machine, gages the opposite end, stopping from the last shoulder previously gaged, and, therefore, gaging the two remaining ones. These two gages do nicely for the machine operations, but it is necessary to furnish still another for the inspection department, like the one shown in Fig. 327. This gage, it will be noticed, checks up the entire length as well as all the other dimensions.

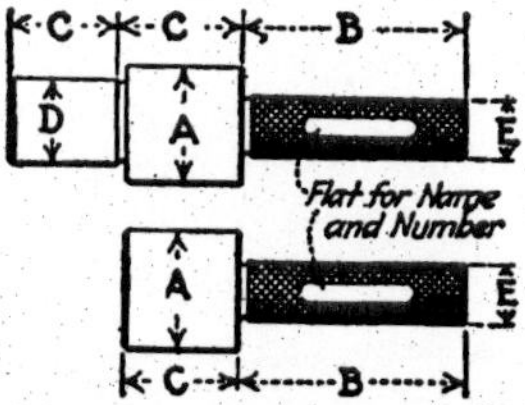

D	A	C	B	E
1/8	3/16	1/8	1 1/4	1/8
3/16	1/4	3/16	1 1/4	1/8
1/4	3/8	1/4	1 1/2	1/4
5/16	7/16	5/16	1 1/2	5/16
3/8	9/16	3/8	1 3/4	5/16
7/16	5/8	7/16	1 3/4	3/8
1/2	3/4	1/2	1 3/4	3/8
9/16	13/16	9/16	2	7/16
5/8	7/8	5/8	2	7/16
3/4	1	3/4	2	1/2
7/8	1 1/8	7/8	2	1/2
1	1 1/4	1	2	1/2

All sizes in inches.

TABLE 37.—SIZES OF DRILL AND COUNTERBORE GAGES FOR FILLISTER-HEAD SCREWS

Fig. 328 illustrates a method of making a profile gage for checking up the lengths for three or more shoulders. Here it is impossible to hook over each shoulder as in the previous example, because the shoulders vary only several thousandths of an inch and the gage is marked with an arrow which represents their respective lengths. A simple but good example of a profile gage is shown in Fig. 330. This style of gage is used for gaging such work as is shown in Fig. 329, and its utility will be readily recognized.

Fig. 331 is a fillister-head screw gage. Gages made up in this style will be found very useful, particularly when there are a great many sizes of a similar part to be gaged, as one takes the place of several which would otherwise be necessary. Its use is, however, restricted to the inspection department.

It may be advisable to make up for the manufacturing department a distinct gage for each size and thus do away with the necessity of having one operator borrowing from the other and the two operators, perhaps, wanting the same gage at the same time, thereby decreasing instead of increasing production. Referring again to the

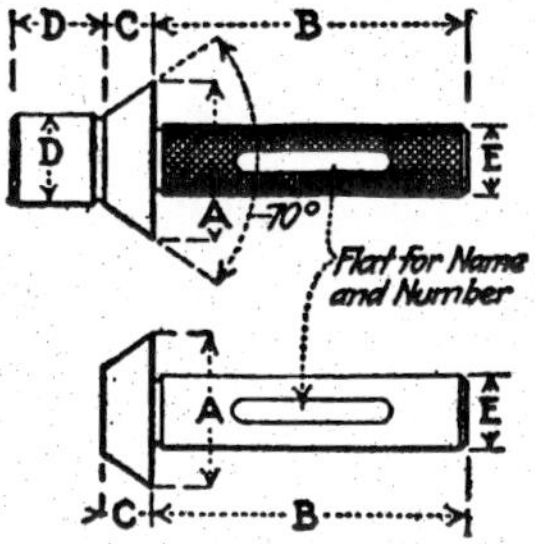

D	A	C	B	E
$\frac{1}{8}$	$\frac{1}{4}$	$\frac{3}{32}$	$1\frac{1}{4}$	$\frac{1}{8}$
$\frac{3}{16}$	$\frac{3}{8}$	$\frac{9}{64}$	$1\frac{1}{4}$	$\frac{1}{8}$
$\frac{1}{4}$	$\frac{15}{32}$	$\frac{5}{32}$	$1\frac{1}{2}$	$\frac{1}{4}$
$\frac{5}{16}$	$\frac{5}{8}$	$\frac{7}{32}$	$1\frac{1}{2}$	$\frac{1}{4}$
$\frac{3}{8}$	$\frac{3}{4}$	$\frac{17}{64}$	$1\frac{3}{4}$	$\frac{5}{16}$
$\frac{7}{16}$	$\frac{13}{16}$	$\frac{17}{64}$	$1\frac{3}{4}$	$\frac{5}{16}$
$\frac{1}{2}$	$\frac{7}{8}$	$\frac{17}{64}$	$1\frac{3}{4}$	$\frac{3}{8}$
$\frac{9}{16}$	1	$\frac{5}{16}$	2	$\frac{3}{8}$
$\frac{5}{8}$	$1\frac{1}{8}$	$\frac{23}{64}$	2	$\frac{1}{2}$
$\frac{3}{4}$	$1\frac{3}{8}$	$\frac{7}{16}$	2	$\frac{1}{2}$

All sizes in inches.

TABLE 38.—SIZES OF DRILL AND COUNTERSINK FOR FLAT-HEAD SCREWS.

illustration, the length *A* represents the diameter and *B* the length of the head of the screw.

RECEIVING GAGES

Receiving gages are well known to those manufacturing small interchangeable parts such as firearms, typewriting, adding machines, etc. Their use, like checking gages, is frequently confined to the inspection department. Figs. 332 and 333 are similar and good examples. They are both made up in three pieces. The lower half acts as the base, while the upper portions are made to the form of the part

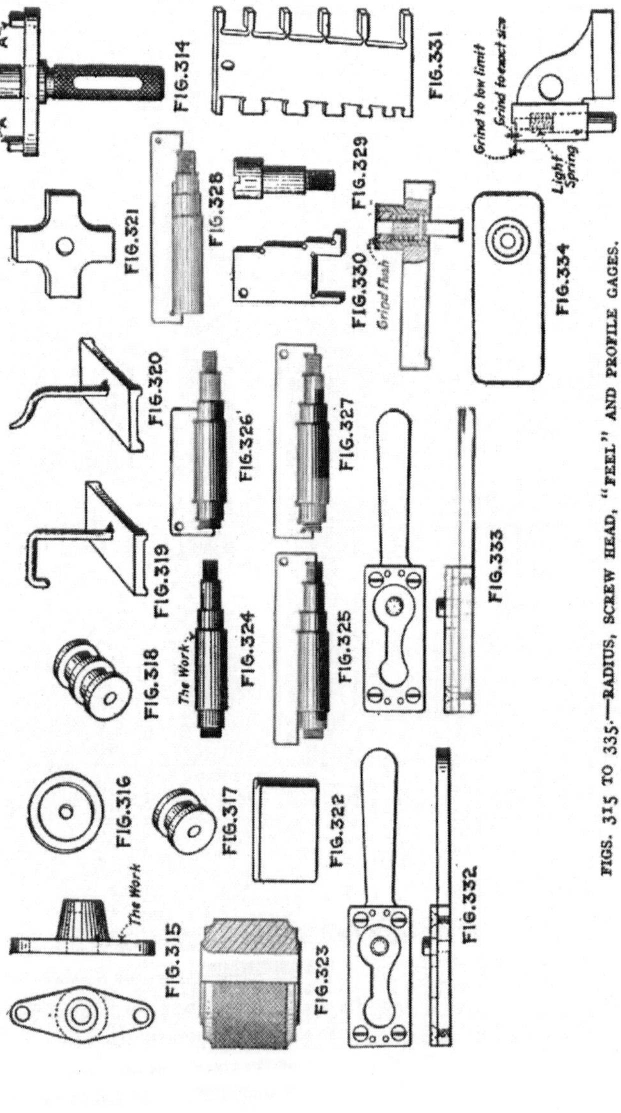

FIGS. 315 TO 335.—RADIUS, SCREW HEAD, "FEEL" AND PROFILE GAGES.

to be gaged, after which they are screwed and doweled, pinned in place, then disassembled, hardened and reassembled to the lower half.

Fig. 333 is an improvement over the type shown in Fig. 332; half of the upper portion projects above the other an amount equal to the thickness of the work to be gaged. This allows the gaging of the first operation independent of the second.

FLUSH PIN GAGES

The old saying that the sense of touch is keener than the eye possibly was instrumental in the designing of a gage to meet the requirements. With the flush pin gage the touch is relied upon more than the eye, and these gages are generally used for gaging such shapes as are difficult to gage with any other form of gage. Figs. 334 and 335 represent the standard method of constructing them, and the illustrations speak for themselves.

A form of gage not very often seen in use is illustrated in Fig. 336. It is a style of limit gage used for checking up the centers of one shaft with another. That part of the gage which is bushed is slipped over the end of one shaft and the graduated end brought in contact with the outer surface of the other shaft, as illustrated.

This end of the gage is hardened and the contact points are lapped to the length required and as the graduations read. The center represents the exact distance and the other points the plus or minus allowance in thousandths of an inch. This gage has many advantages over the indicator type of gage. It is more reliable, less costly and quick to read.

INDICATOR GAGES

Where a very high degree of accuracy is required very delicate gages are necessary, and in such instances the indicator gage will be found in use. Several types of these gages, illustrating the designing and making of them, are shown in Figs. 337 to 343. Fig. 337 is known as an indicator gage; it is used also to test inside or surface work. Fig. 338 illustrates the use of this style of gage for gaging the centers of one shaft with another.

A gage for measuring the thickness of walls of tubing, etc., is shown in Fig. 339, and a compound-lever indicating gage for gaging the diameter of bored holes is shown in Fig. 340. Another form of this style of gage is shown in Fig. 341. Its use is here shown in gaging

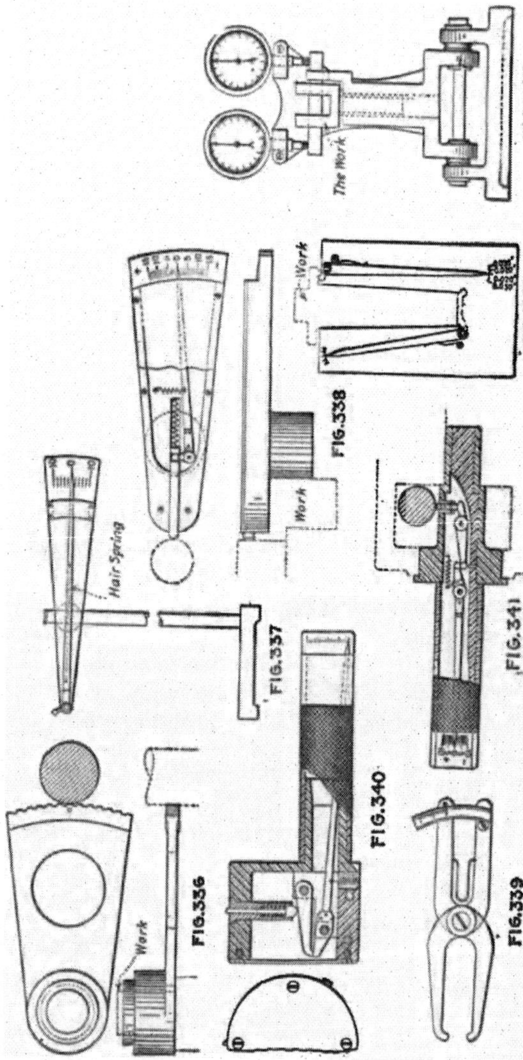

FIGS. 336 TO 343.—DIFFERENT TYPES OF INDICATING GAGES.

one shaft at right angles with another. A profile indicator gage with double indicating points is shown in Fig. 342.

Fig. 343 represents one method of using the dial indicator in gages of the indicating type. The illustration speaks for itself, and as there are so many applications to which this form of gage can be applied, it is unnecessary to say anything further regarding its advantages. When designing indicating gages, it is always advisable to make up a masterplate, this to be used for checking their accuracy from time to time.

POINTS TO BE CONSIDERED BEFORE DESIGNING GAGES

What style of gage is wanted? Snap, plug, ring, indicator, thread, depth, target, profile, reference, checking, height, pin, etc? Are there any standard gages that can be bought instead of making them? Is the gage to be used in the machine shop or inspection department? Is the gage to be used for more than one operation? Will a soft or temporary gage suffice? Have you specified the proper material? Is the gage to be hardened? If so, have you made reference to this on the drawing?

PROFILE GAGES

If the gage is a profile gage, can all shoulders and lengths of parts be gaged, or would it be advisable to make more than one? Let us say, for example, that you have a shouldered shaft, which is made partly on the screw machine and then finished on the engine lathe. Have you made the gages so that they can be used? Is the gage strong enough? Have you provided a hole in the gage so that it may be hung up if necessary? Is the gage too heavy?

RING GAGES

Have you made the gage wide enough? Is the wall of the gage sufficiently heavy to insure the gage remaining true after it has been made? Would a temporary gage made of cast iron do? Have you provided a shoulder around the mouth of the hole for grinding? Have you knurled the gage on the outside? Are the corners chamfered? Is a limit ring gage wanted? If a limit ring gage, are your limits correct? Could not a snap gage be used instead of a ring gage?

Is it necessary that the ring gage take in the radius of the part to be gaged? If so, cut away the gage so that the one using it can readily see if the radius is correct.

PLUG GAGES

If a standard plug gage, is it long enough to go all the way through the hole to be gaged? Is the gage too heavy? Can you lighten it? Would it be cheaper to make it of flat plate stock instead of solid round stock? Can you cheapen the gage by making the handle separate? Would a pin gage do instead of a plug gage? Are the limits such as to require the limit plug gage? Can a limit gage be used to advantage? If a limit plug gage, are your limits correct? Would a temporary soft gage do? Is it necessary for the gage to take in the radius or chamfer at the mouth of the hole? If a large gage, would a four-point plug gage do? Can the plug gage be made to act also as a depth gage?

SNAP GAGES

In making the snap gage, can you use any standard forgings such as are made by forge companies for this purpose? Is the gage to be a limit or standard snap gage? If a limit snap gage, would it not be cheaper or quicker to use the micrometer instead of going to the expense of making a gage? If you have to make a forging have you made it strong enough? Would not a gage made of cast iron with inserted hardened points do? Could you not save making up a number of gages by making a gage with interchangeable jaws? Can the gage be made of standard round-bar or flat-plate stock? Is the gage too heavy; could the body be made of aluminum? If the gage is made of round-bar stock, and to be used on the lathe, have you offset the center to clear the arbor or mandrel? If a limit snap gage, are the limits correct?

INDICATOR GAGES

Is the gage so constructed that the graduations are spaced far enough apart so that they can be plainly read? What provision have you made to prevent injury to the indicating needle? Is the gage too heavy? Is the gage too light? Is the gage strong enough? Are all working parts properly marked and hardened? In case of wear can the gage be adjusted?

THREAD GAGES

Have you provided dirt grooves? Is the gage too heavy or awkward to handle? Is the gage long enough? If a large gage, can it be lightened? What means have you provided for adjusting the gage? In case a female thread gage is wanted, could not the thread micrometer be used instead? If a male thread gage, is the handle knurled and long enough? If a female thread gage, have you knurled the outside? Is it necessary to make up a limit thread gage? If so, are your limits correct?

CHAPTER XIV

THE DESIGNING OF MACHINE VISE JAWS

The many different styles of vises call for a great deal of ingenuity on the part of the designer, and they all, without exception, show that they have been handled by experienced men who realized the necessity of rigid construction, as well as compact and neat design.

A great deal of time has also been spent on the important point of quick-acting vises, although this feature has not been improved upon to any great extent, except by the recent designers.

Many vise manufacturers who still stick to their original design do so probably because of the interchangeability of parts, rather than for any other reason. This means a great deal when it comes to manufacturing and replacing broken parts.

Work held in the vise is the source of much comment between the tool designer and his superior, and frequently the designer is forced to design expensive milling fixtures which, in many cases, are nothing more than a huge vise. The fact that manufacturers supply the vise as part of the equipment of their milling machines demonstrates the general usefulness of the vise. Yearly large sums of money are spent in their vise departments perfecting design and workmanship.

AN ASSORTMENT OF VISES

For general manufacturing purposes the vise illustrated in Fig. 344 is a fairly representative type, and it can be clamped rigidly in almost any position on the miller table. It is this type of vise which the tool designer comes in contact with most when designing jaws for special purposes. Swivel bases, with graduated scales for setting the vise at given angles, are a distinct necessity for toolroom work; but for manufacturing it is best to dispense with an extra movable surface coming between the cutter and the work.

In Figs. 345 to 348 are shown designs of vise jaws suitable for holding small, round stock while milling keyways or similar operations. Two larger shafts may be milled at the same time by providing jaws as illustrated in Fig. 349, when the limits of accuracy are sufficiently wide to allow for slight variation in the size of the bars. Fig. 350 is a good method of holding stock of even size for gang-milling a bar to a number of short lengths.

An almost standard type of jaw is seen in Fig. 351. This kind is used largely in the toolroom for milling the clearance on form cutters, but it demonstrates a method applicable to other work for throwing the job over at an angle. Fig. 352 shows a jaw in which the metal is cut away, as in a die, to hold a forging or casting of irregular shape. The dowel pins *A A* are for lining up and resisting the tendency of the jaws to move sideways, caused by gripping on an irregular form. Without these pins the work of resisting this lateral motion would come entirely on the threaded part of the screws holding the jaw in position. Also by changing the jaws from one vise to another correct alignment is immediately secured.

A similar die-form jaw is shown in Fig. 353. An idea is conveyed of how rigidly a piece may be held by this method for taking a broad heavy cut across a piece of somewhat delicate form. In Fig. 354 the piece is seen located by means of pins. These pins should be made as short as is consistent with proper locating in order to save time in taking out one piece and putting in another.

Another phase of the die-type jaw is shown in Fig. 355. In this case pieces are added, by means of which the location is secured; this is probably a cheaper method in many cases, and also lends itself to easy alteration in the event of slight change in the size or shape of the part being machined. This will appeal to those having had experience on batches of the same casting coming from different foundries.

A very neat method of holding small turned pieces of uniform size, such as screws, pins, etc., is seen in Fig. 356. The number of pieces that a clever operator can mill in a short time by this method is surprising.

The extended jaw, shown in Fig. 357, is a type of very wide application and specially adapted to holding thin flat pieces beyond the capacity of the ordinary range of the vise. The part of the jaw *A* in contact with the side of the piece is cut away to form a gripping edge; this is a very effective method of resisting any lifting tendency. Fig. 358 is also another application of the case when the work is too long to go between the standard jaws of the vise, but is hardly adapted for heavy work.

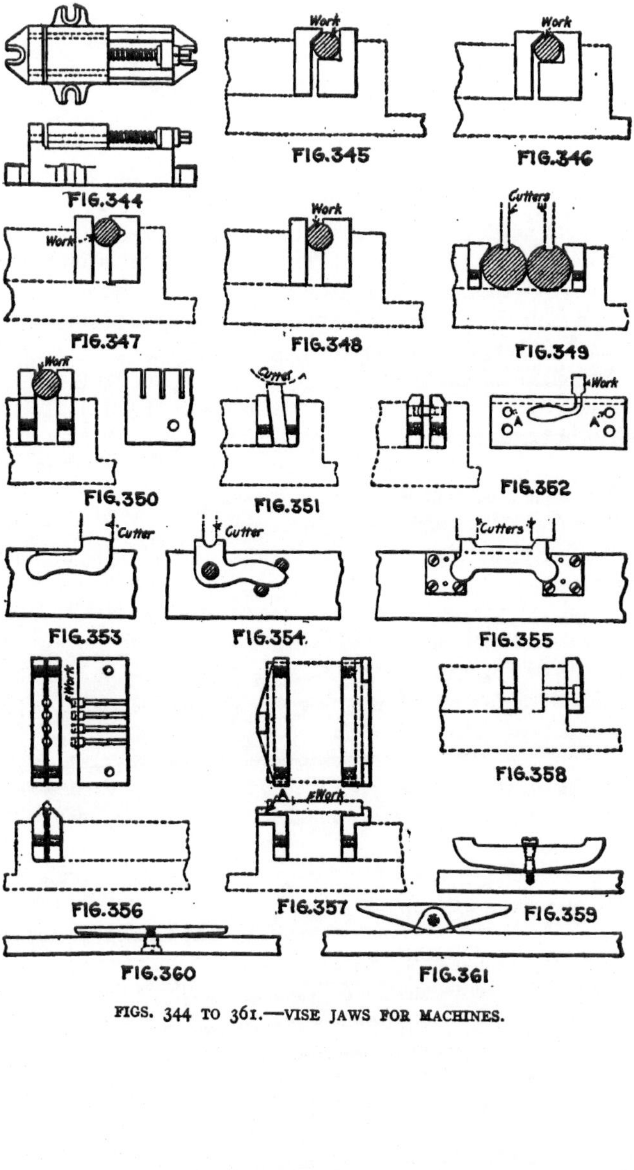

FIGS. 344 TO 361.—VISE JAWS FOR MACHINES.

VISES WITH COMPENSATORS

Three methods of the application of compensating, by which two pieces of unequal size may be held rigidly at one time, or a rough casting gripped on two points, are illustrated in Figs. 359 to 361. In Fig. 359 the compensator is free to move about the retaining screw, while in Fig. 360 the retaining screw is part of the compensator and the movement takes place in the clearance hole in the jaw. In Fig. 361 the compensator is retained in position by a fulcrum pin. It should be noted that in neither case does any pressure come on the

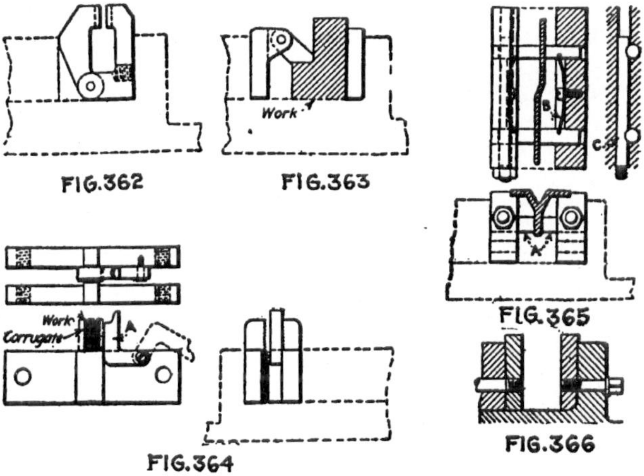

FIGS. 362 TO 366.—VISES WITH SPECIAL JAWS.

retaining device; the pressure is transmitted directly through the compensator to the vise jaw, on which it rests. The examples shown in no way cover the wide range of designs based on this principle.

A type of swivel jaw very suitable for holding small pieces is shown in Fig. 362. It should be observed that the fulcrum pin is in such a position that when the jaw is closing on the piece it has also a slight downward motion just sufficient to counteract the tendency of the piece to rise. Another swivel-piece jaw is seen in Fig. 363. With this device the work may be depended upon to thoroughly bed itself against the bottom and opposite side.

In Fig. 364 is shown the application of a gage *A* for setting the work in the vise; it is attached to the jaw itself. After the work is set correctly the gage may be swung on one side out of the path of the cutter, as indicated by the dotted lines, and as easily brought into position again for setting the next piece.

In Fig. 365 is seen a vise jaw of a more complicated nature, but of very wide application for holding pieces of delicate form, where gripping them in the ordinary way would cause distortion. The hardened pins *A* are held out from the jaw by the flat springs *B*. When it is seen that all these pins are in contact with the work, they are locked in position by the clamping pin *C*, running through the jaw. The vise is then tightened down on the piece in the ordinary way.

It is very necessary that the special jaws described above should be fitted to the vises in a rigid and workmanlike way. Most of the vises on the market are provided with means for doing this. Care should be taken that the hardened jaws are perfectly flat and bed all the way along on their supporting surface. Any burrs on the tapped holes should be removed, and the corners at the bottom of the jaw should be of sufficient radius to make sure they are not touching at the bottom and top only, as shown in Fig. 366.

To get this perfect fit, which is so necessary to eliminate all spring and insure the job being held squarely, hardened jaws should in all cases be ground. Not only is a higher-class finish given to the work held in rigid jaws, but faster feeds and speeds and less broken cutters more than compensate for extra workmanship here.

The material generally used for vise jaws is machine steel, case-hardened. Where the number of the pieces to be machined, or the design of the jaw warrants it, it is advisable to use tool steel, hardened. Cast iron may occasionally be used, but when using this material, it is necessary to design the jaws considerably thicker.

CHAPTER XV

CONSTRUCTION AND USES OF MANDRELS

Mandrels, as differentiated from arbors, are used to hold work, while arbors are used to hold tools, and though some of the examples given in this article may be used occasionally to hold tools they are primarily intended as mandrels.

The designing and making of a mandrel is frequently looked upon as a very simple undertaking, but in order to make one to do the work for which it is wanted, many important points must be considered. One of the first questions a tool designer or maker considers is whether a tapered, straight, expansion or other type of mandrel is needed.

In drill jigs or milling fixtures, it is essential that the work can be inserted or extricated from them in the quickest possible time, and the same requirements apply to the mandrel; that is, it must be so designed as to firmly hold the work, as well as allow it to be quickly placed on or removed.

Before proceeding to the designing of the mandrel, it is important to study the proper designing of the centers as shown in Figs. 367 and 368. As the correct way of making them is clearly illustrated, no explanation is necessary. The use of inserted centers is also illustrated, as these are frequently used in very large mandrels, especially when the body of the mandrel is left soft and a hardened shell placed on it, and they are also extremely useful when there is apt to be any great amount of wear on the centers, because they can be replaced from time to time.

A very important point to be considered, when designing the center, is the oiling, especially when the centers are to be used for a long time for, unless they are properly lubricated, they will soon wear and the accuracy be lost.

COMMON TYPES

Fig. 369 shows what is commonly called a round lathe mandrel, which should be made of tool steel hardened and ground, with a taper of 0.0005 per inch and the small mandrels from 1/4 to 1 inch in diameter should be 0.0005 inch below size at the small end, and figuring the length of a 1/4-in. mandrel at 3 in., the length should increase 1/2 in. for each 1/8 in. of diameter up to 2 1/2 in., but beyond this the length is usually made the same for three sizes and then one inch added to the length for the next group of three.

This style of mandrel is universally used and holds the work through friction, the taper not being enough to distort the hole; however, when work with shoulders is being machined the operator cannot use stops on his machine to set his tools to, because he is unable to press the work on the same distance each time.

Fig. 370 shows a square mandrel made up the same way as the round ones, having the same amount of taper and the advantages and disadvantages are the same in each case.

Fig. 371 illustrates the taper mandrel with an expansion bushing added, which is used when there is apt to be a slight variation in the size of the hole. The bushings are split and ground, taper on the inside to fit the mandrel, and ground straight on the outside diameter. The taper on the mandrel should be 0.0175 per inch and as the bushing is split through it will expand when the tapered mandrel is forced in.

Some tool designers prefer to split the bushings partly through both ways, as shown in Fig. 372, the number of slits varying, as there does not seem to be any standard rule. One tool designer will place as many as eight slits, four each way, in a bushing 1 in. in diameter, while another will use two one way and one the other, but the more general practice seems to be two each way.

SPECIAL EXPANDING MANDRELS

A few designs of expanding mandrels are shown in Figs. 373 to 382 inclusive. Fig. 373 illustrates one which is generally used on a lathe. The shank *A* fits into the spindle and the work to be turned is placed on the split sleeve *B*, which is expanded by the sleeve *C*. The sleeve *B* may be split similar to the expansion bushings shown in Figs. 371 and 372. Attention is called to the pins *D* which hold the bushing in place and keep it from turning. The amount of taper best adapted for the sleeve *C* is from 8 to 10 degrees.

Fig. 374 is another style which differs from Fig. 373 in not having an independent split sleeve. The body *A* is split and expanded by bushing *B*, which is tapered from 10 to 12 degrees and should be a sliding fit on the mandrel *C*. In designing the spring, care should be

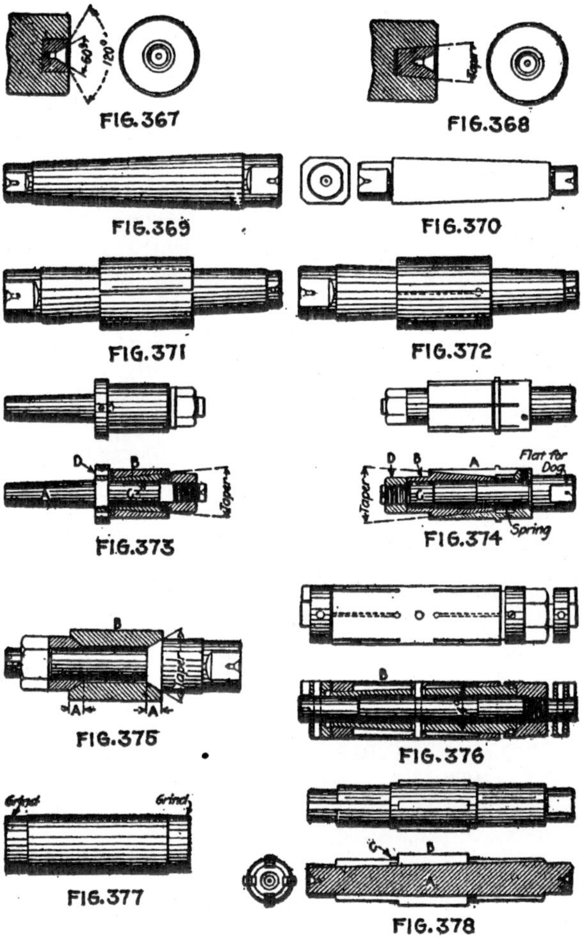

FIGS. 367 TO 378.—ARBORS AND MANDRELS.

taken to make it strong enough to force back the tapered bushing *B*, when the nut *D* is released. This style of arbor allows the operator to place stops on his machine to set his lathe tools to.

Fig. 375 shows another design of the expanding bushing type. It differs from others of this style inasmuch as it expands the bushing from both ends. The illustration speaks for itself. The bushing *B*, if made of cast iron, must be split in six equally spaced points, namely, three one way and three the other, similar to Fig. 372. The amount of taper should be 30 degrees and the length *A* varies. Bushings two inches in diameter by two inches long, to three inches diameter by three inches long and from one-half to three-quarters of an inch thick will give excellent results. The bushings on mandrels of this description are generally made of cast iron, but they may also be made of a good grade of high-carbon steel or tool steel.

Fig. 376 is another style. The bushing *B* is expanded at both ends, similar to the one just shown. The construction of this, though considerably different, gives practically the same results and is used for grinding such work as is shown in Fig. 377, and will give excellent results if made properly. The design shows the driving dogs on both ends, which indicates that the arbor can be reversed.

Fig. 378 shows another, the body *A* of which is slotted with taper slots to receive the blades *B*, which are held in position by sleeve *C*. The slots are tapered ¼ in. to the foot, and an important point to remember in designing mandrels of this sort is to have as many blades as possible.

CHUCKING MANDRELS

Fig. 379 is a mandrel designed for use on chucking machines. The taper *A* is from 16 to 24 deg., and the number of blades should be as many as possible. The body should be made of cast iron and the blades may be made of cold-drawn steel, hardened and ground all over. The rod *B* is used to draw back parts *A A* and expand the blades *C*. This rod should be guided at both ends by means of hardened tool-steel bushings.

Figs. 380 and 381 are similar and illustrate methods of expanding the mandrel, both by means of the pull rod from the rear end of the machine, and also with a screw from the front. The illustrations show them split in three places, but for large mandrels of this description, they may be split in as many places as necessary. It will be

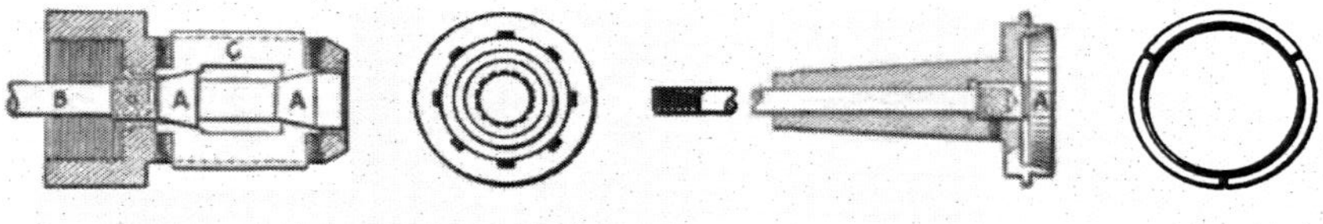

FIG. 379. FIG. 380.

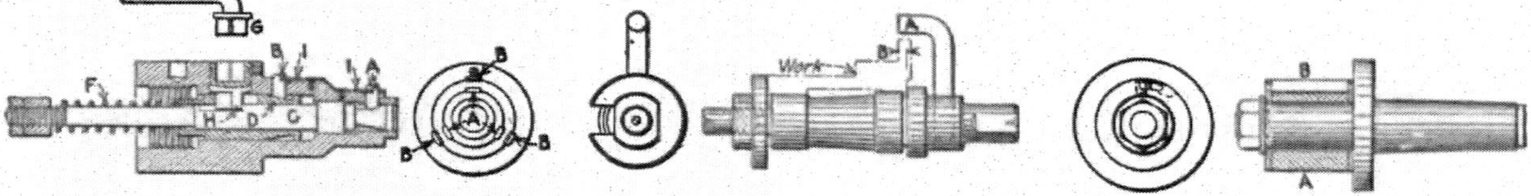

CHUCKING MANDRELS

FIG. 382. FIG. 383. FIG. 384.

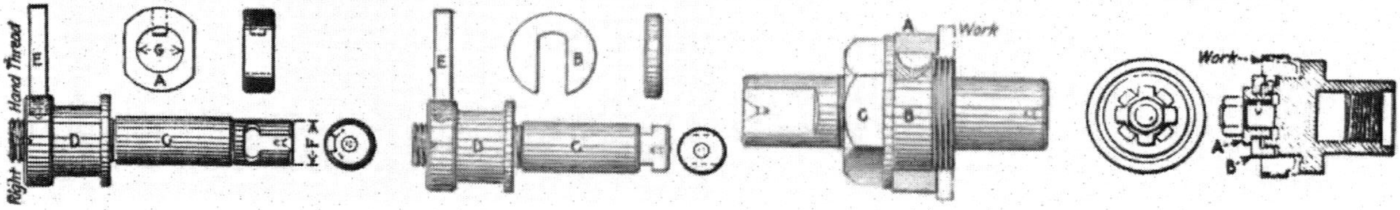

GRINDING MANDRELS

FIG. 385. FIG. 386. FIG. 387. FIG. 388.

FIGS. 379 TO 388.—A VARIETY OF CHUCKING AND GRINDING MANDRELS.

noticed in both cases that the expanding cones *A* are guided, which gives more accurate results than if they were not.

There is one objectionable feature in all expanding mandrels, and that is that dirt or chips are apt to accumulate between the parts, and it is next to impossible to provide dust-proof rings, the only remedy being to keep them properly cleaned while using them.

Another mandrel used on the chucking machine is shown in Fig. 382. This is one of the best known types for chucking up a rough casting so as to turn it true with the inside core, and the one shown was especially made for turning pistons for gasoline engines.

The pins *A* and *B* are forced outward by the rod *C* and the sleeve *DE*, the spring *F* automatically adjusting the rod and sleeve until the front and rear pins have properly seated themselves to the casting. It is well to knurl the outer ends of the pins and also harden them, first making them of the very best obtainable tool steel. To release them, the socket wrench *G* fits into the plug *H*, and as this plug is turned to the right it forces the rod *C* forward and pushes the sleeve *D* back, which in turn allows the pins to release the work. *II* are flat springs which press the pins inward.

SETTING TOOL BY A HARDENED PIN

Fig. 383 is shown to illustrate the advantages of placing a hardened pin *A* on the mandrel for the operator to set his tool to. Let us suppose it is necessary to face the casting from the inside shoulder to an exact measurement *B*, as shown in the illustration, then it can be readily seen that it would be quite difficult to obtain this to any marked degree of accuracy without the pin *A*. The face of this pin should, of course, be hardened and ground.

Fig. 384 is generally called a roller-grip mandrel, and is of the straight type. The work should be a sliding fit on the end *A*, and as soon as the cut begins, the roll *B* grips the work by rolling up the incline, which should be approximately 10 degrees.

Figs. 385 and 386 are grinding mandrels of the straight type. The only difference between the two of them is the lock collar *A* and the slip washer *B*, and they simply illustrate two of the many ways of designing means for obtaining the same result, and it is surprising how quickly work can be removed and placed on mandrels of this description. The work should be a sliding fit on *C*, the nut *D* having a right-hand thread and being driven by the dog, working against the pin *E*, has a tendency to continually bind the work while the arbor is in action. Attention is called to the fit of the lock collar *A* on the end *F* of the arbor shown in Fig. 19. The difference between the dimensions *F* and *G* should be 0.0005 in., which allows plenty of clearance, and the lock collar can be quickly placed on and removed.

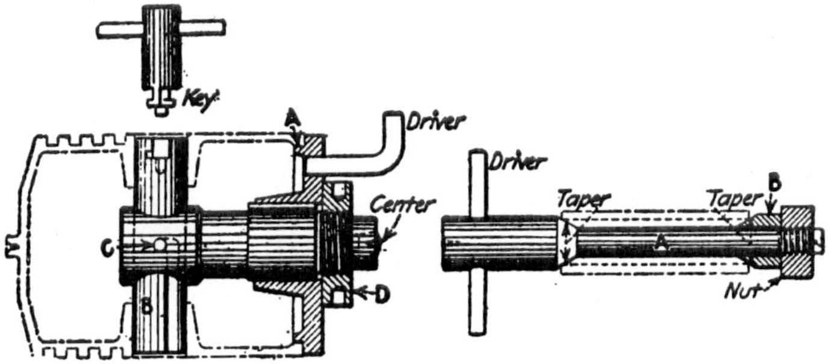

FIGS. 389 AND 390.—MANDREL FOR HOLDING PISTONS.

Fig. 387 is a mandrel, such as used for turning pieces which have been threaded. The cut shows the work in position ready to be faced as well as turned. The parts being threaded right hand natur-

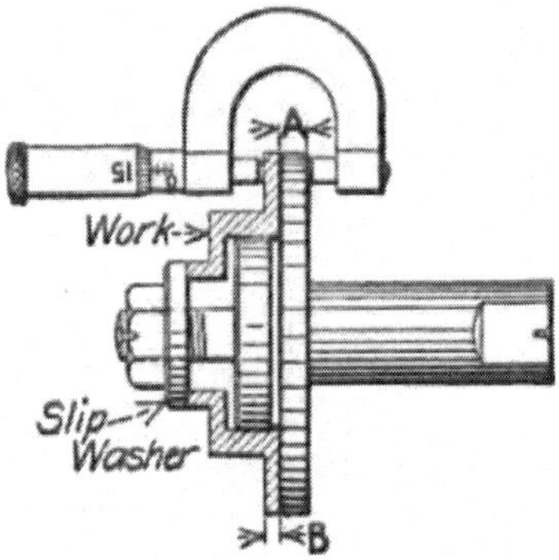

FIG. 391.—MEASURING A FLANGE.

ally tighten against the collar *A* as soon as the cut starts. This collar is keyed to the body *B* and is also a sliding fit, so it can be readily seen that as soon as the nut *C* is loosened the work can be easily removed.

Fig. 388 is shown to illustrate the special clamping washers *A* and *B*. The design speaks for itself, and it will be noticed that the work is such that it will not allow the washer to be removed any other way except from the front.

Fig. 389 is a special design, which is frequently used for grinding pistons and work of a similar nature. It is merely shown to assist the designer or maker, when called upon to make a mandrel for grinding work of this description. The piston in this case is inserted on the flange *A*, after which the pin *B* is put in place. This pin is grooved and when the mandrel is in action it is kept from flying out by the pin *C*, which fits in the groove mentioned. When the pin *B* is in position, the nut *D* is tightened and the work is ready to be ground. To remove the pin *B* a key, similar to the one shown, can be used.

Another style of grinding mandrel for grinding hollow pins or sleeves, is shown in Fig. 390. This gives excellent results, if made of cold-drawn steel, hardened and ground. The taper cone *B* should be a sliding fit on *A*, and the taper should be 60 deg., and it is advisable to make mandrels of this description in duplicate form, so that while one is in use, the operator can unload and load the other.

USING A MICROMETER

It frequently happens that it is necessary to face a flange or boss to a micrometer measurement, and in this case provision should be made for the use of it, and Fig. 391 illustrates how this may be done.

In all cases of this description it is advisable to make the flange *A* of the mandrel a micrometer measurement of even figure, such as 0.375, 0.500 or 0.625 in., which will greatly assist the operator in getting quick results and a correct reading.

INDEX